Statistically Speaking

A Dictionary of Quotations

About the Compilers

Carl C Gaither was born in 1944 in San Antonio, Texas. He has conducted research work for the Texas Department of Corrections and for the Louisiana Department of Corrections. Additionally he has worked as an Operations Research Analyst for the past ten years. He received his undergraduate degree (Psychology) from the University of Hawaii and has graduate degrees from McNeese State University (Psychology), North East Louisiana University (Criminal Justice), and the University of Southwestern Louisiana (Mathematical Statistics).

Alma E Cavazos-Gaither was born in 1955 in San Juan, Texas. She has worked in quality control, material control, and as a bilingual data collector. She received her associate degree (Telecommunications) from Central Texas College.

Statistically Speaking
A Dictionary of Quotations

Selected and Arranged by

Carl C Gaither
and
Alma E Cavazos-Gaither

Institute of Physics Publishing
Bristol and Philadelphia

IOP Publishing Ltd has attempted to trace the copyright holders of all the quotations reproduced in this publication and apologizes to copyright holders if permission to publish in this form has not been obtained.

British Library Cataloguing-in-Publication Data
A catalogue record for this book is available from the British Library.

ISBN 0 7503 0401 4

Library of Congress Cataloging-in-Publication Data
Gaither, Carl C., 1944–
     Statistically speaking : a dictionary of quotations / selected and
arranged by Carl C. Gaither and Alma E. Cavazos-Gaither.
          p.   cm.
     Includes bibliographical references (p. –) and index.
     ISBN 0-7503-0401-4 (alk. paper)
     1. Probabilities– –Quotations, maxims, etc.   2. Mathematical
statistics– –Quotations, maxims, etc.   I. Cavazos-Gaither, Alma E.,
1955–   II. Title.
QA273.G3124 1996
519.5– –dc20
                                                            96-44176
                                                            CIP

Published by Institute of Physics Publishing, wholly owned by The Institute of Physics, London

Institute of Physics Publishing, Techno House, Redcliffe Way, Bristol BS1 6NX, UK
US Editorial Office: Institute of Physics Publishing, Suite 1035, The Public Ledger Building, 150 South Independence Mall West, Philadelphia, PA 19106, USA

Typeset in TEX using the IOP Bookmaker Macros
Printed in Great Britain by J W Arrowsmith Ltd, Bristol

We respectfully dedicate this book to our parents

Mr and Mrs C C Gaither
and
Ms M Cavazos

CONTENTS

CONTENTS

PREFACE

Statistically Speaking is a book of quotations. It has, for the first time, brought together in one easily accessible form the best expressed thoughts that are especially illuminating and pertinent to the disciplines of probability and statistics. Some of the quotations are profound, others are wise, some are witty, but none are frivolous. Quotations from the most famous men and women lie in good company with those from unknown wits. You may not find all the quoted 'jewels' that exist, but we are certain that you will find a great number of them here. We believe that Benjamin Franklin was correct when he said that "Nothing gives an author so much pleasure as to find his work respectfully quoted...".

Statistically Speaking is also an aid for the individual who loves to quote – and to quote correctly. "Always verify your quotations" was advice given to Dean John William Bourgen, then fellow of Oriel College, by Dr Martin Joseph Routh. That advice was given over 150 years ago and is still true today. Frequently, books on quotations will have subtle changes to the quotation, changes to punctuation, slight changes to the wording, even misleading information in the attribution, so that the compiler will know if someone used a quotation from 'their' book. We attempted to verify each and every one of the quotations in this book to ensure that they are correct.

The attributions give the fullest possible information that we could find to help you pinpoint the quotation in its appropriate context or discover more quotations in the original source. Judicial opinions and speeches include, when possible, the date of the opinion or speech. We assure the reader that not one of the quotations in this book was created by us.

In summary, *Statistically Speaking* is a book that has many uses. You can:

- Identify the author of a quotation.
- Identify the source of the quotation.
- Check the precise wording of a quotation.
- Discover what an individual has said on a subject.
- Find sayings by other individuals on the same subject.

xi

How to Use This Book

1. A quotation for a given subject may be found by looking for that subject in the alphabetical arrangement of the book itself. To illustrate, if a quotation on likelihood is wanted, you will find nine quotations listed under the heading likelihood. The arrangement of quotations in this book under each subject heading constitutes a collective composition that incorporates the sayings of a range of people.

2. To find all the quotations pertaining to a subject and the individuals quoted use the SUBJECT BY AUTHOR INDEX. This index will help guide you to the specific statement that is sought. A brief extract of each quotation is included in this index.

3. If you recall the name appearing in the attribution or if you wish to read all of an individual author's contributions that are included in this book then you will want to use the AUTHOR BY SUBJECT INDEX. Here the authors are listed alphabetically along with their quotations. The birth and death dates are provided for the authors whenever we could determine them. When we could not find the information we included a (–).

Thanks

It is never superfluous to say thanks where thanks are due. First, I thank my stepdaughter Maritza Marie Cavazos for her assistance in tracking down incomplete citations, looking for books in the libraries, and helping to sort the piles of correspondence generated in obtaining permissions. Next, we thank the following libraries for allowing us to use their resources: the main library and the science library of The University of Richmond; the main library of the Virginia Commonwealth University; the medical library of the Virginia Commonwealth Medical School; the main library and the science library of Baylor University; the main library of the University of Mary-Hardin Baylor; the main library of the Central Texas College; the main library, the physics-math-astronomy library, and the human resource library of the University of Texas at Austin.

Additionally, we would like to thank each of the publishers who provided permission to use the quotations. We made a very serious attempt to contact the publishers for permission to use the quotations. Letters were written to each publisher or agent for which we could find an address. A follow-up letter was sent to those who did not respond to our first letter. If no response was received we then assumed a calculated risk and incorporated the quotation. In no way did we use a quotation without attempting to obtain prior approval.

Carl Gaither
Alma Cavazos-Gaither

ACTUARY

Analytical and graphical treatment of statistics is employed by the economist, the philanthropist, the business expert, the actuary, and even the physician, with the most surprising valuable results . . .

Karpansky, L.
High School Education
Chapter 6 (p. 134)

Someone once asked an accountant, a mathematician, an engineer, a statistician and an actuary how much 2 plus 2 was. The accountant said "4". The mathematician said "It all depends on your number base." The engineer took out his slide-rule and said "approximately 3.99". The statistician consulted his tables and said, "I am 95% confident that it lies between 3.95 and 4.05." The actuary said "What do you want it to add up to?"

Unknown

Actuaries are funny people. Even when they are wrong, they are right. I told an actuary to go to the back of the queue. He immediately came back and said that he couldn't—there was already someone there.

Unknown

An insurance company is like an automobile going down the road at high speed. The managing director has his hands on the wheel, the marketing director has his foot on the accelerator. The finance director is heaving with all his might on the hand-brake and the actuary is in the back screaming directions from a map he has just made by looking out of the rear window.

Unknown

ANALYSIS

Not even the most subtle and skilled analysis can overcome completely the unreliability of basic data.

Allen, R.G.D.
Statistics for Economists
Chapter I (p. 14)

The technical analysis of any large collection of data is a task for a highly trained and expensive man who knows the mathematical theory of statistics inside and out. Otherwise the outcome is likely to be a collection of drawings—quartered pies, cute little battleships, and tapering rows of sturdy soldiers in diversified uniforms—interesting enough in a colored Sunday supplement, but hardly the sort of thing from which to draw reliable inferences.

Bell, Eric T.
Mathematics: Queen and Servant of Science (p. 383)

He was in *Logick*, a great *Critick*,
Profoundly skill'd in Analytick;
He could distinguish and divide
A hair 'twixt south and south-west side.

Butler, Samuel
Hudibras
Part I, Canto I, l. 65

The repetition of a catchword can hold analysis in fetters for fifty years and more.

Cardozo, Benjamin N.
Harvard Law Review
Mr. Justice Holmes
Volume 44, Number 5, March 1931 (p. 689)

Murphy's Laws of Analysis. (1) In any collection of data, the figures that are obviously correct contain errors. (2) It is customary for a decimal to be misplaced. (3) An error that can creep into a calculation, will. Also, it will always be in the direction that will cause the most damage to the calculation.

Deakly, G.C.
Quoted in Paul Dickson's
The Official Rules (M-126)

The mere fact of naming an object tends to give definiteness to our conception of it—we have then a sign that at once calls up in our minds the distinctive qualities which mark out for us that particular object from all others.

Eliot, George
The George Eliot Letters
Volume II (p. 251)

It is not the first use but the tiresome repetition of inadequate catchwords which I am observing—phrases which originally were contributions, but which, by their very felicity, delay further analysis for fifty years.

Holmes, O.W., Jr.
Collected Legal Papers (pp. 230–1)

I have seen too much not to know that the impression of a woman may be more valuable than the conclusion of an analytical reasoner . . .

Holmes, Sherlock
in Arthur Conan Doyle's
The Complete Sherlock Holmes
The Man with the Twisted Lip

. . . be wary of analysts that try to quantify the unquantifiable.

Keeney, Ralph
Raiffa, Howard
Decisions with Multiple Objectives: Preferences and Value Trade-Offs (p. 12)

But to argue, without analysis of the instances, from the mere fact that a given event has a frequency of 10 percent in the thousand instances under observation, or even in a million instances, that . . . it is likely to have a frequency near to 1/10 in a further set of observations, is . . . hardly an argument at all.

Keynes, John Maynard
Treatise on Probability
Chapter XXXIII (p. 407)

An intelligence that, at a given instant, could comprehend all the forces by which nature is animated and the respective situation of the beings that make it up, if moreover it were vast enough to submit these data to analysis, would encompass in the same formula the movements of the greatest bodies of the universe and those of the lightest atoms. For such an intelligence nothing would be uncertain, and the future, like the past, would be open to its eyes.

Laplace, Pierre-Simon
A Philosophical Essay on Probabilities (p. 2)

Sweet Analytics, 'tis thou hast ravish'd me . . .

Marlowe, Christopher
Christopher Marlowe's Doctor Faustus
Scene 1

. . . the habit of analysis has a tendency to wear away the feelings.

Mill, John Stuart
Autobiography
V (p. 116)

The very excellence of analysis . . . tends to weaken and undermine whatever is the result of prejudice; that it enables us mentally to separate ideas which have only casually clung together . . .

Mill, John Stuart
Autobiography
V (p. 116)

As in Mathematics, so in Natural Philosophy, the Investigation of difficult Things by the Method of Analysis, ought ever to precede the Method of Composition. This Analysis consists in making Experiments and Observations, and in drawing general Conclusions from them by Induction, and admitting of no Objections against the Conclusions but such as are taken from Experiments, or other certain Truths.

Newton, Sir Isaac
Opticks
Book III, Part I

Analysis, Cross-reference analysis,
   The age of analysis.
Psychological, philosophical, poetic analysis.
   Not the event, but the picturing of the event.

Sherman, Susan
With Anger/With Love
The Fourth Wall
Stanza 2

"Our company's president built a financial empire on the 50–50 future theory," the manager told a new employee.

"Oh, you mean he used probability analysis to forecast and make business decisions?"

"No, nothing like that," the manager answered. "I mean he believes that every $50 raise he doesn't give you increases future profits by the same amount."

Thomsett, Michael C.
The Little Black Book of Business Statistics (p. 74)

If data analysis is to be well done, much of it must be a matter of judgment, and "theory", whether statistical or non-statistical, will have to guide, not command.

Tukey, John W.
Annals of Mathematical Statistics
The Future of Data Analysis
Volume 33, Number 1, March 1962 (p. 10)

It always helps to know the answer when you are working toward the solution of a problem.

Unknown

It requires a very unusual mind to undertake the analysis of the obvious.

Whitehead, Alfred North
Science and the Modern World (p. 4)

AVERAGE

If at first you don't succeed, you are running about average.

Alderson, M.H.
Quoted in Paul Dickson's
The Official Explanations (p. A-4)

In respect of honour and dishonour, the observance of the mean is Greatness of Soul, the excess a sort of Vanity, as it may be called, and the deficiency, Smallness of Soul.

Aristotle
The Nicomachean Ethics
Book II, Chapter 7

. . . but they are more hysterical than the average because they have the opportunity their constituents lack, of shouting in public.

Atherton, Gertrude
Senator North
Book II, VII

The average intelligence is always shallow, and in electric climates very excitable.

Atherton, Gertrude
Senator North
Book II, IX

There must be such a thing as a child with average ability, but you can't find a parent who will acknowledge that it is his child . . .

Bailey, Thomas D.
Wall Street Journal
Notable and Quotable
December 17, 1962 (p. 16)

Another very frequent application of mathematics to biology is the use of averages which, in medicine and physiology, leads, so to speak, necessarily to error . . . By destroying the biological character of phenomena, the use of *averages* in physiology and medicine usually gives only apparent accuracy to the results.

Bernard, Claude
An Introduction to the Study of Experimental Medicine (p. 134)

Chemical averages are also often used. If we collect a man's urine during twenty-four hours and mix all this urine to analyze the average, we get an analysis of a urine which simply does not exist; for urine, when fasting, is different from urine during digestion. A startling instance of this kind was invented by a physiologist who took urine from a railroad station urinal where people of all nations passed, and who believed he could thus present an analysis of *average* European urine!

Bernard, Claude
An Introduction to the Study of Experimental Medicine (pp. 134–5)

About the hardest thing a phellow kan do, iz tew spark two girls at onest, and preserve a good average.

Billings, Josh
Old Probability: Perhaps Rain—Perhaps Not
May 1870

Great numbers and the averages resulting from them, such as we always obtain in measuring social phenomena, have great inertia.

Bowley, Arthur L.
Elements of Statistics
Part I, Chapter I (p. 8)

Of itself an arithmetic average is more likely to conceal than to disclose important facts; it is the nature of an abbreviation, and is often an excuse for laziness.

Bowley, Arthur L.
The Mathematical Gazette
Volume 12, Number 77, July 1925
#319 (p. 421)

I abhor averages. I like the individual case. A man may have six meals one day and none the next, making an average of three meals per day, but that is not a good way to live.

Brandies, Louis D.
Quoted in Alpheus T. Mason's
Brandies: A Free Man's Life (p. 145)

Have shaving too entailed upon their chins,—
A daily plague, which in the aggregate
May average on the whole with parturition.

Byron, Lord
Don Juan
Canto XIV, 23–4

The best way of increasing the [average] intelligence of scientists would
be to reduce their number.

Carrel, Alexis
Man the Unknown
Chapter 2, 4 (p. 49)

The concept of *average* was developed in the Rhodian laws as to the
distribution of losses in maritime risks.

Cohen, Morris R.
Journal of the American Statistical Association
The Statistical View of Nature
Volume 31, Number 194, June 1936 (p. 328)

. . . the criminal intellect, which its own professed students perpetually
misread, because they persist in trying to reconcile it with the average
intellect of average men instead of identifying it as a horrible wonder
apart . . .

Dickens, Charles
The Work of Charles Dickens
The Mystery of Edwin Drood
XX

The plain man is the basic clod
From which we grow the demigod;
And the average man is curled
The hero stuff that rules the world.

Foss, Sam Walter
Back Country Poems
Memorial Day
Stanza 2

True, the *average* rate for the year as a whole, though on the high side, is
not too bad, but that is like assuring the nonswimmer that he can safely
walk across a river because its *average* depth is only 4 feet.

Freidman, Martin
Newsweek
Irresponsible Monetary Policy
January 10, 1972 (p. 57)

Unfortunately, the average of one generation need not be the average of the next.

Froude, James Anthony
Short Studies on Great Subjects
The Science of History (p. 26)

There is no medium at sea. You are either dead sick or ravenous, and we, not excluding the two boys were the latter.

Froude, James Anthony
Short Studies on Great Subjects
A Fortnight in Kerry (p. 195)

We have to consider the million, not the units; the average, not the exceptions.

Froude, James Anthony
Short Studies on Great Subjects
On Progress (p. 261)

My friends at Rhodes made me so. I cost as much as sixteen gold gods of average size.

Froude, James Anthony
Short Studies on Great Subjects
Lucian (p. 225)

The knowledge of an average value is a meager piece of information.

Galton, Francis
Natural Inheritance
Scheme of Distribution and of Frequency (p. 35)

It is difficult to understand why statisticians commonly limit their enquiries to Averages, and do not revel in more comprehensive views. Their souls seem as dull to the charm of variety as that of the native of one of our flat English counties, whose retrospect of Switzerland was that, if its mountains could be thrown into its lakes, two nuisances would be got rid of at once. An average is but a solitary fact, whereas if a single other fact be added to it, an entire Normal Scheme, which nearly corresponds to the observed one, starts potentially into existence.

Galton, Francis
Natural Inheritance
The Charms of Statistics (p. 62)

But though to visit the sins of the fathers upon the children may be a morality good enough for divinities, it is scorned by average human nature; and it therefore does not mend the matter.

Hardy, Thomas
Tess of the d'Urbervilles
XI

Give me a man that is capable of a devotion to anything, rather than a cold, calculating average of all the virtues!

Harte, Francis Bret
Two Men of Sandy Bar
Act IV (p. 425)

If a man stands with his left foot on a hot stove and his right foot in a refrigerator, the statistician would say that, on the average, he's comfortable.

Heller, Walter
in Harry Hopkins'
The Numbers Game: The Bland Totalitarianism
Chapter 12, Faithful Partners
Counter Attack (p. 270)

They had on average, about a quarter of a suit of clothes and one shoe apiece. One chap was sitting on the floor of the aisle, looking as if he were working a hard sum in arithmetic. He was trying very solemn, to pull a lady's number two shoe on a number nine foot.

Henry, O.
Tales of O. Henry
Holding Up a Train

But an average, which was what I meant to speak about, is one of the most extraordinary subjects of observation and study.

Holmes, O.W.
The Autocrat of the Breakfast Table
Chapter 6

On the average, bunting with a man on first loses a lot of runs. On the average, it doesn't increase the probability of scoring at least one run in the inning.

Hooke, Robert
Quoted in J.M. Tanur's
Statistics: A Guide to the Unknown
Statistics, Sports, and Some Other Things

There is a *mean* in things, fixed limits on either side of which right living cannot get a foothold.

Horace
The Complete Works of Horace
The Golden Mean (p. 6)

The average man believes a thing first, and then searches for proof to bolster his opinion.

Hubbard, Elbert
The Philistine: A Periodical of Protest
Volume XI, July 1900 (p. 36)

Fertilize and bokanovskify—in other words, multiply by seventy-two—and you get an average of nearly eleven thousand brothers and sisters in a hundred and fifty two batches of identical twins, all within two years of the same age.

Huxley, Aldous
Brave New World (p. 7)

. . . public opinion, a vulgar, impertinent, anonymous tyrant who deliberately makes life unpleasant for anyone who is not content to be the average man.

Inge, William Ralph
Outspoken Essays
Our Present Discontents (p. 9)

The average man is rich enough when he has a little more than he has got, and not till then.

Inge, William Ralph
Outspoken Essays
Patriotism (pp. 38–9)

Such is the past career, present condition, and certain future of the Middle American. There are as many above him as below him, and especially as many below him as above him.

Jacobs, Joseph
American Magazine
The Middle American
Volume 63, March 1907

"Pardon me for staring," said Milo, after he had been staring for some time, "but I've never seen half a child before."

"It's .58 to be precise," replied the child from the left side of his mouth (which happened to be the only side of his mouth).

"I beg your pardon?" said Milo.

"It's .58," he repeated; "it's a little bit *more* than a half."
. . .

"Oh, we're just the average family," he said thoughtfully; "mother, father, and 2.58 children—and, as I explained, I'm the .58."

Juster, Norton
The Phantom Tollbooth (pp. 195–6)

"But averages aren't real," objected Milo, "they're just imaginary."

"That may be so," he agreed, "but they're also very useful at times. For instance, if you didn't have any money at all, but you happened to be with four other people who had ten dollars apiece, then you'd each have an average of eight dollars. Isn't that right?"

Juster, Norton
The Phantom Tollbooth (p. 196)

. . . 'hitting the target', for centuries the principal military skill, is henceforth to be left to the law of averages.

Keegan, John
The Face of Battle (p. 307)

One need not accept Shaw's own estimate of his intellectual equipment to see that the doctor's remark cut through a confusion in which psychologists and sociologists flounder. Frequently they make no distinction between what is "normal" and what is "usual", "average", or "statistically probable".

Krutch, Joseph Wood
Human Nature and the Human Condition
Chapter 5 (p. 75)

. . . the question "How many legs does a normal man have?" should be answered by finding a statistical average. And since some men have only one leg, or none, this would lead inevitably to the conclusion that a "normal" man is equipped with one and some fraction legs.

Krutch, Joseph Wood
Human Nature and the Human Condition
Chapter 5 (p. 76)

All very old men have splendid educations; all men who apparently know nothing else have thorough classical educations; nobody has an average education.

Leacock, Stephen
Literary Lapses
A Manual of Education (p. 127)

Dear Sir,—We beg to acknowledge your letter of application and cheque for fifteen dollars. After careful comparison of your case with the average modern standard, we are pleased to accept you as a first-class risk.

Leacock, Stephen
Literary Lapses
Insurance up to Date (p. 158)

What does this mean for The Average Man?

Lieber, Lillian R.
The Education of T.C. MITS (p. 71)

In former times, when the hazards of sea voyages were much more serious than they are today, when ships buffeted by storms threw a portion of their cargo overboard, it was recognized that those whose goods were sacrificed had a claim in equity to indemnification at the expense of those whose goods were safely delivered. The value of the lost goods was paid for by agreement between all of those whose merchandise had been in the same ship. This sea damage to cargo in transit was known as 'havaria' and the word came naturally to be applied to the compensation money which each individual was called upon to pay. From this Latin word derives our modern word *average*.

Moroney, M.J.
Facts from Figures
On the Average (p. 34)

A want of the habit of observing and an inveterate habit of taking averages are each of them often equally misleading.

Nightingale, Florence
Notes on Nursing
Chapter XIII

The average American is just like the child in the family.

Nixon, Richard M.
The New York Times
Statement from Pre-Election Interviews with Nixon Outlining 2nd Term Plans
Page 20, Column 8
November 10, 1972

For, I ask, what is man in Nature? A cypher compared with the Infinite, an All compared with Nothing, a mean between nothing and all.

Pascal, Blaise
Pascal's Pensées
Section I, 43

. . . it suggests *Haverie*—average, you know . . .

Pynchon, Thomas
Gravity's Rainbow (p. 207)

l'homme moyen
[the average man]

Quetelet, Adolphe
A Treatise on Man and the Development of His Faculties (p. 100)

Make sure that the real average is what you are dealing with.

Redfield, Roy A.
Factors of Growth in a Law Practice (p. 170)

Great minds discuss ideas, average minds discuss events, small minds discuss people.

Rickover, H.G.
The Saturday Evening Post
The World of the Uneducated
November 28, 1959 (p. 59)

Scientific laws, when we have reason to think them accurate, are different in form from the common-sense rules which have exceptions: they are always, at least in physics, either differential equations, or statistical averages.

Russell, Bertrand A.
The Analysis of Matter
Chapter XIX (p. 191)

The Normal is the good smile in a child's eyes—all right. It is also the dead stare in a million adults. It both sustains and kills—like a God. It is the Ordinary made beautiful; it is also the Average made lethal.

Shaffer, Peter
Two Plays by Peter Shaffer
Equus
Act I, Scene 19

Nerissa. They are as sick that surfeit with too much as they that starve with nothing. It is no mean happiness therefore, to be seated in the mean: superfluity comes sooner by white hairs, but competency lives longer.

Shakespeare, William
The Complete Works of William Shakespeare
Merchant of Venice
Act I, Scene 2, l. 5

It is a well-known statistical paradox that the average age of women over forty is under forty . . .

Slonim, Morris James
Sampling (p. 26)

"You can't fight the law of averages," Grover said, "you can't fight the curve."

Snood, Grover
Quoted in Thomas Pynchon's
Slow Learner
The Secret Integration (p. 142)

Ask a ferryman or a toll-keeper how many visitors come through daily on an average, and with an appearance of great intellectual discomfort he assures you the number varies so much, "Some days it's a lot, and some days only a few, there isn't exactly an average".

Stamp, Josiah
Some Economic Factors in Modern Life
Chapter VII (p. 253)

Sir,—In your issue of December 31 you quoted Mr. B.S. Morris as saying that many people are disturbed that about half the children in the country are below the average in reading ability. This is only one of many similarly disturbing facts. About half the church steeples in the country are below average height; about half our coal scuttles below average capacity, and about half our babies below average weight. The only remedy would seem to be to repeal the law of averages.

Stewart, Alan
The Times
Averages
Monday, January 4, 1954 (p. 7)

GUIL: The law of averages, if I have got this right, means that if six monkeys were thrown up in the air for long enough they would land on their tails about as often as they would land on their —

Stoppard, Tom
Rosencrantz and Guildenstern are Dead
Act One (p. 13)

The equanimity of your average tosser of coins depends upon a law, or rather a tendency, or let us say a probability, or at any rate a mathematically calculable chance, which ensures that he will not upset himself by losing too much nor upset his opponent by winning too often.

Stoppard, Tom
Rosencrantz and Guildenstern are Dead
Act One (p. 19)

Expectation in the general sense may be considered as a kind of average.
The Encyclopaedia Britannica
11th Edition
Probability

The wise student hears of the Tao and practices it diligently. The average student hears of the Tao and gives it thought now and again.

Tsu, Lao
Tao Te Ching (Forty-one)

The only very marked difference between the average civilized man and the average savage is that the one is guilded and the other painted.

Twain, Mark
Mark Twain Laughing
1904, #370 (p. 98)

I was very young in those days, exceedingly young, marvelously young, younger than I am now, younger than I shall ever be again, by hundreds of years. I worked every night from eleven or twelve until broad day in the morning, and as I did 200,000 words in the sixty days, the average was more than 3,000 words a day—nothing for Sir Walter Scott, nothing for Louis Stevenson, nothing for plenty of other people, but quite handsome for me. In 1897, when we were living in Tedworth Square, London, and I was writing the book called *Following the Equator*, my average was 1,800 words a day; here in Florence (1904) my average seems to be 1,400 words per sitting of four or five hours.

Twain, Mark
The Autobiography of Mark Twain
Chapter 29

The average man's a coward . . . The average man don't like trouble and danger.

Twain, Mark
Huckleberry Finn
XXII

In medio fortissimus ibis.
[Always choose the middle road.]

Unknown

If we start with the assumption, grounded on experience, that there is uniformity in this average, and so long as this is secured to us, we can afford to be perfectly indifferent to the fate, as regards causation, of the individuals which compose the average.

Venn, J.
The Logic of Chance
Chance, Causation, and Design
Section 4 (p. 239)

Why do we resort to averages at all?

Venn, J.
Journal of the Royal Statistical Society
On the Nature and Uses of Averages
Volume 54, 1891 (p. 429)

How can a single introduction of our own [*average*], and that a fictitious one, possibly take the place of the many values which were actually given to us? And the answer surely is, that it can *not* possibly do so; the one thing cannot take the place of the other for purposes in general, but only for this or that specific purpose.

Venn, J.
Journal of the Royal Statistical Society
On the Nature and Uses of Averages
Volume 54, 1891 (p. 430)

We have seen that man in general, one with another, or (as it is expressed) on the average, does not live above two-and-twenty years . . .

Voltaire
Philosophical Dictionary
Miscellany

Cecily: Mr. Moncrieff and I are engaged to be married, Lady Bracknell.

Lady Bracknell [*with a shiver, crossing to the sofa and sitting down*]: I do not know whether there is anything peculiarly exciting in the air of this particular part of Hertfordshire, but the number of engagements that go on seems to me considerably above the proper average that statistics have laid down for our guidance.

Wilde, Oscar
The Importance of Being Earnest: A Trivial Comedy for Serious People
Act III (p. 118)

BAYESIAN

I am not altogether facetious in suggesting that, while non-Bayesians should make it clear in their writings whether they are *non-Bayesian Orthodox* or *non-Bayesian Fisherian*, Bayesians should also take care to distinguish their various denominations of *Bayesian Epistemologists, Bayesian Orthodox*, and *Bayesian Savage*.

Bartlett, M.S.
Journal of the Royal Statistical Society
Discussion on Professor Pratt's Paper (p. 197)

I believe that assumptions are useful to state in statistical practice, because they impose a discipline on the user. Once a full set of assumptions is stated, the conclusion should follow. (Actually, only a Bayesian analysis can meet this standard, but that's another topic for another time.)

Kadane, Joseph
Statistical Science
Comment
Volume 1, Number 1, February 1986 (p. 12)

I have seen the collective noun for statisticians cited as "a variance of statisticians". I prefer "a skewer of statisticians". There might also be some specialized terminology for Bayesians, but I have not seen any.

Luchenbruch, Peter
Unknown source

. . . there are at least 46,656 varieties of Bayesians.

Wang, Chamont
Sense and Nonsense of Statistical Inference (p. 158)

CAUSE AND EFFECT

Give me to learn each secret cause;
Let number's figure motion's laws
Revealed before me stand;
These to great Nature's secret apply,
And round the Globe, and through the sky,
Disclose her working hand.

Akenside, Mark
The Poetical Works of Mark Akenside and John Dyer
Hymn to Science in Works of the English Poets (p. 357)

The universal cause is one thing, a particular cause another. An effect can be haphazard with respect to the plan of the second, but not of the first. For an effect is not taken out of the scope of one particular cause save by another particular cause which prevents it, as when wood dowsed with water will not catch fire. The first cause, however, cannot have a random effect in its own order, since all particular causes are comprehended in its causality. When an effect does escape from a system of particular causality, we speak of it as fortuitous or a chance happening . . .

Aquinas, Thomas
Summa Theologiae
Part I
Question 22. God's Providence
Article 2. Is everything subject to divine Providence?

Thus all the action of men must necessarily be referred to seven causes: chance, nature, compulsion, habit, reason, anger, and desire.

Aristotle
The Art of Rhetoric
Book I, Chapter X

Only a few look at causes, and trace them to their effects.

Arthur, T.S.
Ten Nights in a Bar Room and What I Saw There
Night the Fifth

The law of cause and effect does not hide in the realm of the unexpected when intelligent beings go looking for it.

Atherton, Gertrude
Senator North
Book II, XXI

In the series of things those which follow are always aptly fitted to those which have gone before . . .

Aurelius, Marcus
The Meditations of the Emperor Antonius Marcus Aurelius
Book IV, Section 45

The end of our foundation is the knowledge of causes, and secret motions of things; and the enlarging of the bounds of human empire, to the effecting of all things possible.

Bacon, Francis
New Atlantis (p. 288)

. . . the present contains nothing more than the past, and what is found in the effect was already in the cause.

Bergson, Henri
Creative Evolution (p. 17)

First causes are outside the realm of science; they forever escape us in the sciences of living as well as in those of inorganic bodies.

Bernard, Claude
An Introduction to the Study of Experimental Medicine (p. 66)

Every effect becomes a cause.

Buddhist Maxim

The Causes of events are ever more interesting than the events themselves.

Cicero
Epistolae ad atticum
Book IX, Section 5

The most important events are often determined by very trivial causes.

Cicero
Orationes Philippicae
V

We know the effects of many things, but the causes of few; experience, therefore, is a surer guide than imagination, and inquiry than conjecture.

Colton, Charles Caleb
Lacon: or many things in a few words (p. 111)

There is no result in nature without a cause; understand the cause and you will have no need of the experiment.

da Vinci, Leonardo
The Notebooks of Leonardo da Vinci
Philosophy (p. 64)

I understand that to be CAUSE OF ITSELF (*causa sui*) whose essence involves existence and whose nature cannot be conceived unless existing.

de Spinoza, Benedict
Ethics
Concerning God
Definition I

III. From a given determined cause an effect follows of necessity, and on the other hand, if no determined cause is granted, it is impossible that an effect should follow.

de Spinoza, Benedict
Ethics
Concerning God
Axiom III

. . . that all men are born ignorant of the causes of things, and that all have a desire of acquiring what is useful; . . .

de Spinoza, Benedict
Ethics
Concerning God
Appendix

But great things spring from causalities.

Disraeli, Benjamin
Sybil or the Two Nations
Book V, III (p. 345)

Happy the man, who studying Nature's laws,
Through known effects can trace the secret cause—
His mind, possessing in a quiet state,
Fearless of fortune and resigned to fate.

Dryden, John
The Poetical Works of Dryden
Translation of Virgil
The Second Book of the Georgics, l. 701

Cause and effect are two sides of one fact.

Emerson, Ralph Waldo
Essays
Circles

Cause and effect, means and ends, seed and fruit, cannot be severed; for the effect already blooms in the cause; the end preexists in the means, the fruit in the seed.

Emerson, Ralph Waldo
Essays
Compensation

Do not clutch at sensual sweetness until it is ripe on the slow tree of cause and effect.

Emerson, Ralph Waldo
Essays
Prudence

Cause and effect, the chancellors of God.

Emerson, Ralph Waldo
Essays
Self-Reliance

Some play at chess, some at cards, some at the Stock Exchange. I prefer to play at Cause and Effect.

Emerson, Ralph Waldo
The Journals of Ralph Waldo Emerson (p. 234)

Shallow men believe in luck, believe in circumstances . . . Strong men believe in cause and effect.

Emerson, Ralph Waldo
Conduct of Life
Worship (pp. 191–2)

Primary causes are unknown to us; but are subject to simple and constant laws, which may be discovered by observation, the study of them being the object of natural philosophy.

Fourier, Jean Baptiste Joseph
Analytical Theory of Heat
Preliminary Discourse

Every effect has its cause.

Froude, James Anthony
Short Studies on Great Subjects
Calvinism (p. 12)

Causation depends on an extraordinary turning of reality at a particular instant such that one event transmutes into another.

Heise, David R.
Causal Analysis (p. 6)

But he who, blind to universal laws,
Sees but effects, unconscious of the cause,—

Holmes, O.W.
The Complete Poetical Works of Oliver Wendell Holmes
A Metrical Essay

. . . you have erred perhaps in attempting to put colour and life into each of your statements, instead of confining yourself to the task of placing upon record that severe reasoning from cause to effect which is really the only notable feature about the thing.

Holmes, Sherlock
in Arthur Conan Doyle's
The Complete Sherlock Holmes
The Adventure of the Copper Beeches

"A coincidence! Here is one of the three men who we had named as possible actors in this drama, and he meets a violent death during the very hours when we know that the drama was being enacted. The odds are enormous against its being a coincidence. No figures could express them. No, my dear Watson, the two events are connected—*must* be connected. It is for us to find the connection."

Holmes, Sherlock
in Arthur Conan Doyle's
The Complete Sherlock Holmes
The Adventure of the Second Stain

In a word, then, every effect is a distinct event from its cause.

Hume, David
An Enquiry Concerning Human Understanding
Section IV (p. 28)

From causes which appear *similar* we expect similar effects. This is the sum of all our experimental conclusions.

Hume, David
An Enquiry Concerning Human Understanding
Section IV (p. 35)

It is universally allowed that nothing exists without a cause of its existence, and that chance, when strictly examined, is a mere negative word, and means not any real power which has anywhere a being in nature.

Hume, David
An Enquiry Concerning Human Understanding
Section VIII (p. 99)

All effects follow not with like certainty from their supposed causes.

Hume, David
An Enquiry Concerning Human Understanding
Section X (p. 115)

Here is a billiard ball lying on the table, and another ball moving toward it with rapidity. They strike; the ball which was formerly at rest now acquires a motion. This is as perfect an instance of the relations of cause and effect as any which we know either by sensation or reflection.

Hume, David
An Enquiry Concerning Human Understanding
An Abstract of A Treatise of Human Nature (pp. 186–7)

As in the night all cats are gray, so in the darkness of metaphysical criticism all causes are obscure.

James, William
The Principles of Psychology
V

With earth's first clay they did the last man knead,
And there of the last harvest sowed the seed.
And the first morning of creation wrote
What the last dawn of reckoning shall read.

James, William
Unitarian Review and Religious Magazine
The Dilemma of Determinism
Volume XXII, Number 3, September 1884

Pure mathematics can never deal with the possibility, that is to say, with the possibility of an intuition answering to the conceptions of the things. Hence it cannot touch the question of cause and effect, and consequently, all the finality there observed must always be regarded simply as formal, and never as a physical end.

Kant, Immanuel
Philosophical Writings
The Critique of Judgment
Critique of Teleological Judgment
63, fn

Causes are often disproportionate to effects.

Lee, Hannah Farnham
The Log Cabin, or, The World before You
Part the Second

Man is a creature who searches for causes; he could be named the cause-searcher within the hierarchy of minds.

Lichtenberg, Georg
Lichtenberg: Aphorisms & Letters
Aphorisms (p. 62)

The truth that every fact which has a beginning has a cause, is co-extensive with human experience.

Mill, John Stuart
System of Logic
Book III, V, 1

Before the effect one believes in other causes than after the effect.

Nietzsche, Friedrich
The Complete Works of Friedrich Nietzsche
The Joyful Wisdom, III, Number 217

The cause is hidden, but the enfeebling power of the fountain is well known.

Ovid
Metamorphoses
IV, l. 287

Rem Viderunt, Causomnon Viderunt.
[They saw the thing, but not the cause.]

Pascal, Blaise
The Thoughts of Blaise Pascal
On the Necessity of the Wager
#235

Sutch as the cause of every thing is, sutch wilbe the effect.

Pettie, George
A Petite Pallace of Pettie His Pleasure
Volume I
Germanicus and Agrippina

On the assumption that all happens by Cause, it is easy to discover the nearest determinants of any particular act or state to trace it plainly to them.

Plotinus
The Six Enneads
Third Ennead
First Tractate, Fate, 1

We must rather seek for a cause, for every event whether probable or improbable must have some cause.

Polybius
The Histories
Book II, 38.5

If the law of the relation of effect and cause does not exist, then the non-existence of cause will follow also from non-existence of effect. Non-existence of effect is not instrumental towards the non-existence of cause; but non-existence of cause is instrumental towards non-existence of effect.

Prakash, Satya
Founders of Sciences in Ancient India (p. 322)

Sublata causa, tollitur effectus.
[The cause being taken away, the effect is removed.]

Proverb, Latin

Post hoc, ergo propter hoc.
[After this, therefore because of this.]

Proverb, Latin

Every Effect Presupposes some Cause.

Rohault, Jacques
Rohault's System of Natural Philosophy
Volume I, Part I, Chapter 5, 6

. . . for no more by the law of reason than by the law of nature can anything occur without a cause.

Rousseau, Jean Jacques
The Social Contract
Book II, Chapter 4

But we are not likely to find science returning to the crude form of causality believed in by Fijians and philosophers of which the type is "lightning causes thunder".

Russell, Bertrand A.
The Analysis of Matter
Chapter XI (p. 102)

. . . and now remains
That we find out the cause of this effect,
Or rather say, the cause of this defect,
For this effect defective comes by cause.

Shakespeare, William
The Complete Works of William Shakespeare
Hamlet, Prince of Denmark
Act II, Scene 2, l. 100

There is occasions and causes why and wherefore in all things.

Shakespeare, William
The Complete Works of William Shakespeare
The Life of King Henry the Fifth
Act V, Scene 1, l. 3

It is the cause, it is the cause, my soul—
Let me not name it to you, you chaste stars!—
It is the cause.

Shakespeare, William
The Complete Works of William Shakespeare
Othello, The Moor of Venice
Act V, Scene 2, l. 1

Thou art the cause, and most accursed effect.

Shakespeare, William
The Complete Works of William Shakespeare
The Tragedy of King Richard the Third
Act I, Scene 2, l. 120

Looking for long-term causes of things is like ascribing motor accidents to the existence of the internal combustion engine.

Taylor, J.P.
London Review Books 3(1)

Wherefore by their fruits ye shall know them.

The Bible
Matthew 7:20

The combination of phenomena is beyond the grasp of the human intellect. But the impulse to seek cause is innate in the soul of man. And the human intellect, with no inkling of the immense variety and complexity of circumstances conditioning a phenomenon, any one of which may be separately conceived as the cause of it, snatches at the first and most easily understood approximation, and says here is the cause.

Tolstoy, Leo
War and Peace
Book XII, Chapter 1

Everything can be a "that"; everything can be a "this". One man cannot see things as another sees them . . . Therefore it is said "'That' comes from 'this' and 'this' comes from 'that'"—which means "that" and "this" give birth to one another.

Tsu, Chuang
Inner Chapters (p. 29)

I am not a heretic; I do believe in causality.

Unknown

The cause is the same with a Barmter (a Barometer I suppose she meant, if she meant anything), which has a great Effect on the Weather. Say rather the Weather has a great Effect on it.

Unknown
Adventures of Sylvia Hughes
Written by herself, 48

Happy is he who has been able to learn the causes of things, . . .

Virgil
Quoted in James Lonsdale's
The Works of Virgil
The Georgics
II, l. 489

CERTAINTY

... if a man will begin with certainties he shall end in doubts, but if he will be content to begin with doubts, he shall end in certainties.

Bacon, Francis
The Advancement of Learning
First Book (p. 41)

Oh! let us never doubt
What nobody is sure about!

Belloc, Hilaire
More Beasts for Worse Children
The Microbe

There is one thing certain, namely that we can have nothing certain; and therefore it is not certain that we can have nothing certain.

Butler, Samuel
Samuel Butler's Note-Books (p. 195)

... we're not sure, we can't be sure. Otherwise, there would be a solution; at least one could get oneself taken seriously.

Camus, Albert
The Fall (p. 74)

Sometimes the probability in favor of a generalization is enormous, but the infinite probability of certainty is never reached.

Dampier-Whetham, William
Science and the Human Mind
Chapter X

It was not a PERHAPS; it was a certainty.

Froude, James Anthony
Short Studies on Great Subjects
Times of Erasmus, Desderius and Luther (p. 47)

"Certainty," Father Newman insists, is the same in kind wherever and by whomsoever it is experienced. The gravely and cautiously formed conclusion of the scientific investigator, and the determination of the school-girl that the weather is going to be fine, do not differ from each other so far as they are acts of the mind.

Froude, James Anthony
Short Studies on Great Subjects
The Grammar of Assent (p. 105)

If one thing is more certain than another—which is extremely doubtful—

Galsworthy, John
End of the Chapter
Maid in Waiting
Chapter 13

He is a fool who leaves certainties for uncertainties.

Hesiod
Fragments
Frag 18 (p. 278)
Quoted by Plutarch
Moralia
Section 505D

We can be absolutely certain only about things we do not understand.

Hoffer, Eric
The True Believer
Part 3, Chapter XII, Section 57 (p. 79)

Heads I win, Tails you lose.

Holmes, O.W.
The Professor at the Breakfast Table (p. 223)

But certainty generally is illusion, and repose is not the destiny of man.

Holmes, O.W., Jr.
Harvard Law Review
The Path of the Law
Volume 10, Number 7, February 25, 1897 (p. 466)

Certitude is not the test of certainty. We have been cock-sure of many things that were not so.

Holmes, O.W., Jr.
Harvard Law Review
Natural Law
Volume 32, Number 1, November 1918 (p. 40)

. . . we can know nothing . . . *for certain* . . .

Jeans, James Hopwood
The New Background of Science (p. 58)

When speculation has done its worst, two and two still make four.

Johnson, Samuel
The Idler

Yet I shall not deny that the number of phenomena which are happily explained by a given hypothesis may be so great that it may be taken as morally certain.

Leibniz, Gottfried Wilhelm
Leibniz: Philosophical Papers and Letters
Volume I
On the Elements of Natural Science (p. 347)

. . . the highest probability amounts not to certainty, without which there can be no true knowledge.

Locke, John
An Essay Concerning Human Understanding
Book IV, III, 14

As mathematical and absolute certainty is seldom to be attained in human affairs, reasoning and public utility require that judges and all mankind in forming their opinion of the truth of facts should be regulated by the superior number of probabilities on the one side or the other.

Mansfield, Lord
Quoted in Francis Wellman's
The Art of Cross-Examination (p. 168)

I must have certainty. Give it to me; or I will kill you when next I catch you asleep.

Shaw, George Bernard
Back to Methuselah
Act I
In the Beginning

Not a resemblance, but a certainty.

Shakespeare, William
The Complete Works of William Shakespeare
Measure for Measure
Act IV, Scene 2, l. 203

All predictions are statistical, but some predictions have such a high probability that one tends to regard them as certain.

Walker, Marshall
The Nature of Scientific Thought (p. 70)

Heads I win, Tails you lose.
O.W. Holmes – (See p. 31)

CHANCE

A substantial portion of the lecture was devoted to risks . . . He emphasized that one in a million is a very remote risk.

Abelson, Philip H.
Science
Editorial
4 February 1994

A Frenchman named Chamfort, who should have known better, once said that chance was a nickname for Providence.

Ambler, Eric
A Coffin for Dimitrios (p. 1)

In all such beings chance occurs, not in the sense that everything about them occurs by chance, but that in each of them there is room for chance and this very fact is a sign that they are subject to someone's rule.

Aquinas, Thomas
Summa Theologiae
Part I
Question 103. God's Government taken as a Whole
Article 5. Whether all things are subject to God's government

Clearly none of the traditional sciences concerns itself with the accidental.

Aristotle
Metaphysics
Book XI, Chapter VIII

To begin with, then, we note that some things follow upon others uniformly or generally, and it is evidently not such things that we attribute to chance or luck.

Aristotle
The Physics
Book II, Chapter V

. . . chance is excluded from natural events, and whatever applies everywhere and to all cases is not to be ascribed to chance.

Aristotle
On the Heavens
Book II, Chapter VIII

. . . rational action is merely a question of calculating the chances.

Aron, Raymond
The Opium of the Intellectuals
Chapter VI
The Illusion of Necessity (p. 165)

Chance is the fool's name for fate.

Astaire, Fred
The movie *The Gay Divorcee*

Games of chance are probably as old as the human desire to get something for nothing; but their mathematical implications were appreciated only after Fermat and Pascal in 1654 reduced chance to law.

Bell, Eric T.
The Development of Mathematics (p. 154)

Every night and every morn
Some to misery are born;
Every morn and every night
Some are born to sweet delight.

Blake, William
The Complete Writings of William Blake
Poems from the Pickering Manuscript
Auguries of Innocence, l. 119–21

Of all the gin joints in all the towns in all the world, she walks into mine!

Bogart, Humphrey
The movie *Casablanca*

Can there be laws of chance? The answer, it would seem. should be negative, since chance is in fact defined as the characteristic of the phenomena which follow no law, phenomena whose causes are too complex to permit prediction.

Borel, Emile
Probabilities and Life
Introduction (p. 1)

The conception of chance enters into the very first steps of scientific activity in virtue of the fact that no observation is absolutely correct. I think chance is a more fundamental conception than causality; for whether in a concrete case, a cause–effect relationship holds or not can only be judged by applying the laws of chance to the observation.

Born, Max
Natural Philosophy of Cause and Chance (p. 47)

What we still designate as chance, merely depends on a concatenation of circumstances, the internal connection and the final causes of which we have as yet been unable to unravel.

Buchner, Ludwig
Force and Matter
The Fitness of Things in Nature (p. 179)

It must always be remembered that man's body is what it is through having been molded into its present shape by the chances and changes of an immense time . . .

Butler, Samuel
Erewhon
Chapter XXII

We see but a part, and being thus unable to generalize human conduct, except very roughly, we deny that it is subject to any fixed laws at all, and ascribe much both of a man's character and actions to chance, or luck, or fortune . . .

Butler, Samuel
Erewhon
Chapter XXIII

Quoth She: "I've heard old cunning *Stagers*
Say, Fools for *Arguments* use wagers."

Butler, Samuel
Hudibras
The Second Part
Canto I, verses 298–9

Quelqu'un disait que la providence strat le nom de baptême du hasard . . .
[Chance is a nickname of Providence.]

Chamfort, Sebastien Roch
Maximes et pensées
Ib. 62

"Proof!" he cried. "Good God! the man is looking for proof! Why, of course, the chances are twenty to one that it has *nothing* to do with them. But what else can we do? Don't you see we must either follow one wild possibility or else go home to bed."

Chesterson, Gilbert Keith
The Father Brown Omnibus
The Innocence of Father Brown
The Blue Cross

Surely nothing is so at variance with reason and stability as chance.

Cicero
Cicero: De Senectute, De Amicitia, De Divinatione
De Divinatione
II, vii

. . . but things that happen by chance cannot be certain.

Cicero
Cicero: De Senectute, De Amicitia, De Divinatione
De Divinatione
II, ix

As in the game of billiards, the balls are constantly producing effects from mere chance, which the most skillful player could neither execute nor foresee, but which, when they *do* happen, serve mainly to teach him how much he has still to learn . . .

Colton, Charles Caleb
Lacon: or many things in a few words (p. 345)

One has to be extraordinarily lucky, in our society, to meet one nymphomaniac in a lifetime.

Comfort, Alex
Darwin and the Naked Lady: Discursive Essays on Biology and Art
The Rape of Andromeda (p. 87)

A fool must now and then be right, by chance.

Cowper, William
Cowper: Poetical Works
Conversation, l. 96

Chance is the only source of true novelty.

Crick, Francis Harry Compton
Life Itself (p. 82)

When the game of hazard is broken up, he who loses remains sorrowful
. . .

Dante
The Divine Comedy of Dante Alighiere
Purgatory
Canto 6, l. 1–2

When we look at the plants and bushes clothing an entangled bank, we are tempted to attribute their proportional numbers and kinds to what we call chance. But how false a view this is!

Darwin, Charles
The Origin of Species
Chapter III

I am inclined to look at everything as resulting from designed laws, with the details, whether good or bad, left to the working out of what we may call chance.

Darwin, Charles
The Life and Letters of Charles Darwin
Volume II
C. Darwin to Asa Gray
May 22nd [1860] (p. 105)

. . . some of the Problems about Chance having a great appearance of Simplicity, the Mind is easily drawn into a belief, that their Solution may be attained by the mere Strength of natural good Sense; Which generally providing otherwise and the Mistakes occasioned thereby being not unfrequent. 'Tis presumed that a Book of this Kind, which teaches to distinguish Truth from what seems nearly to resemble it, will be looked upon as a help to good reasoning.

de Moivre, Abraham
The Doctrine of Chances (p. ii)

There are many People in the World who are prepossessed with an Opinion, that the Doctrine of Chances has a Tendency to promote Play; but they soon will be undeceived . . . this Doctrine is so far from encouraging Play, that it is rather a Guard against it, by setting in a clear Light, the Advantages and Disadvantages of those Games wherein Chance is concerned.

de Moivre, Abraham
The Doctrine of Chances
Dedication

Nothing can come into being from that which is not, or pass away into what is not.

Democritus
in Diogenes Laterius'
Lives of Eminent Philosophers
Chapter IX

She hadn't a Chinaman's chance.

Disney, Dorothy
Crimson Friday (p. 206)

Be juster, Heav'n: such virtue punish'd thus,
Will make us think that Chance rules all above,
And shuffles, with a random hand, the Lots
Which Man is forc'd to draw.

Dryden, John
The Poetical Works of Dryden
All For Love
Act V

There was once a brainy baboon,
Who always breathed down a bassoon,
For he said "It appears
That in billions of years
I shall certainly hit on a tune."

Eddington, Sir Arthur Stanley
New Pathways in Science
Chapter III, Section IV
The End of the World (p. 62)

In our scientific expectation we have grown antipodes. You believe in
God playing and I in perfect laws in the world of things existing as real
objects, which I try to grasp in a wildly speculative way.

Einstein, Albert
Letter to Max Born
7 November 1944

Value depends upon price and price upon chance and caprice.

Eldridge, Paul
Maxims for a Modern Man
1855

Great Jove!
What shall I say? that thou from Heaven look'st down
Upon mankind, or have they rashly formed
A vain opinion, deeming that the race
Of gods exists, though fortune governs all?

Euripides
The Plays of Euripides
Hecuba, l. 486

A general is a man who takes chances. Mostly he takes a fifty–fifty
chance; if he happens to win three times in succession he is considered
a great general.

Fermi, Enrico
Quoted in Leo Szilard's
Leo Szilard: His Version of the Facts (p. 147)

There are fifty ways which I may go after I leave my door. The odds are forty-nine to one against my taking any particular way that can be mentioned, yet a person says that he saw me go that way and not another, his evidence is accepted without difficulty, and the fact is taken to be proved.

Froude, James Anthony
Short Studies on Great Subjects
The Grammar of Assent (p. 109)

It's all chance, but we can't stop now.

Galsworthy, John
End of the Chapter
Maid in Waiting, Chapter 28

The whimsical effects of chance in producing stable results are common enough. Tangled strings variously twitched, soon get themselves into tight knots. Rubbish thrown down a sink is pretty sure in time to choke the pipe; no one bit may be so large as its bore, but several bits in their numerous chance encounters will at length so come into collision as to wedge themselves into a sort of arch across the tube, and effectively plug it.

Galton, Francis
Natural Inheritance
Organic Stability (p. 21)

He had left off being a perfectionist then, when he discovered that not promptly kept appointments; not a house circumspectly clean, not even membership in Onwentsa, or the Lake Forest Golf and Country Club, or the Lawyer's Club, not power—not *anything*—cleared you through the terrifying office of chance; that it is chance and not perfection that rules the world.

Guest, Judith
Ordinary People
Chapter 11

The odds are still about five to one against hitting the right combination, but that is better than no odds at all.

Harrison, Harry
Astounding
The Mothballed Spaceship

If there is a 50–50 chance that something can go wrong, then 9 times out of 10 it will.

Harvey, Paul
Paul Harvey News, 1979

. . . chance, that is, an infinite number of events, with respect to which our ignorance will not permit us to perceive their causes, and the chain that connects them together. Now, this chance has a greater share in our education than is imagined. It is this that places certain objects before us and, in consequence of this, occasions more happy ideas, and sometimes leads us to the greatest discoveries . . .

Helvetius, C.A.
On Mind
Essay III, Chapter I (p. 196)

If chance be generally acknowledged to be the author of most discoveries in almost all the arts, and if in speculative sciences its power be less sensibly perceived, it is not perhaps less real . . .

Helvetius, C.A.
On Mind
Essay III, Chapter IV (p. 221)

. . . it is well to bear in mind that chances rule men, and not men chances.

Herodotus
The History of Herodotus
Volume II, Book VII, 49

Roll dem bones . . .

Heyward, DuBose
Carolina Chansons: Legends of the Low Country
Gamesters All

Then let a man now face the foe and perish or be saved: such is the intercourse of war.

Homer
The Iliad of Homer
Book XVII, 226

Though there be no such thing as *Chance* in the world, our ignorance of the real cause of any event has the same influence on the understanding, and begets a like species of belief or opinion.

Hume, David
An Enquiry Concerning Human Understanding
Section VI (p. 37)

Nothing was ever said with uncommon felicity, but by the cooperation of chance; and therefore, wit, as well as valor must be content to share its honors with fortune.

Johnson, Samuel
The Yale Edition of the Works of Samuel Johnson
The Idler and the Adventurer
Idler No. 58

Caput aut navia
[Heads or Tails]

Latin Expression

I shot an arrow into the air,
It fell to earth I know not where,
For so swift it flew, the sight
Could not follow it in its flight.

Longfellow, Henry Wadsworth
The Poems of Longfellow
The Arrow and the Song

"I should estimate," this scientist was supposed to have said, "that there is one chance in ten nothing will happen with the bomb, and one chance in a hundred that it will ignite the atmosphere."

Masters, Dexter
The Accident (p. 16)

But then I was reading in the paper just the other day about one of them saying there wasn't more than one chance in God-knows-what, a trillion maybe, that these Bikini bombs could blow up the world. I said to myself, this seems pretty safe odds. But then I said to myself, hey! how come any odds at all? Who's running this show anyway? I sort of get to wondering every once in a while whether anybody knows the middle and the end of what's going on as well as the beginning.

Masters, Dexter
The Accident (p. 382)

. . . that power which erring men call Chance.

Milton, John
Poetical Works of John Milton
Volume II
Comus
l. 587

. . . Chance governs all.

Milton, John
Paradise Lost
Book II, l. 910

No conqueror believes in chance.

Nietzsche, Friedrich
The Complete Works of Friedrich Nietzsche
The Joys of Wisdom, III, Number 258

There must be *chance* in the midst of design; by which we mean, that events which are not designed, necessarily arise from the pursuit of events which are designed. One man travelling to York, meets another man travelling to London.

Paley, William
Natural Theology
Volume II, Goodness of the Deity (p. 186)

The *appearance of chance* will always bear a proportion to the ignorance of the observer.

Paley, William
Natural Theology
Volume II, Goodness of the Deity (p. 186)

Cleopatra's nose—had it been shorter, the whole face of the earth would have been changed.

Pascal, Blaise
Pascal's Pensées
Section I, 93

A game is on, at the other end of this infinite distance, and heads or tails will turn up. What will you wager?

Pascal, Blaise
Pascal's Pensées
Section I, 223

In the field of experimentation, chance favors only the prepared mind.

Pasteur, Louis
in René Dubos'
Louis Pasteur: Free Lance of Science (p. 101)

Nick the Greek's Law of Life. All things considered, life is 9 to 5 against.

Peers, John
1001 Logical Laws (p. 50)

Crito. But you see, Socrates, that the opinion of the many must be regarded, for what is now happening shows that they can do the greatest evil to any one who has lost their good opinion.

Socrates. I only wish it were so, Crito; and that the many could do the greatest evil; for then they would be able to do the greatest good—and what a fine thing this would be! But in reality they can do neither, for they cannot make a man either wise or foolish; and whatever they do is the result of chance.

Plato
Crito
44

. . . in human affairs chance is almost everything.

Plato
Laws
Book IV, 709

Athenian Stranger. They say that the greatest and fairest things are the work of nature and of chance, the lesser of art, which, receiving from nature the greater and primeval creations, molds and fashions all those lesser works which are generally termed artificial.

Plato
Laws
Book X, 889

The lover of intellect and knowledge ought to explore causes of intelligent nature first of all, and, secondly, of those things which, being moved by others, are compelled to move others. And this is what we too must do. Both kinds of causes should be acknowledged by us, but a distinction should be made between those which are endowed with mind and are the workers of things fair and good, and those which are deprived of intelligence and always produce chance effects without order or design.

Plato
Timaeus
46

But from outside there is no knowing which is true. From outside, there is a five-tenths chance that the cat's alive.

But a cat can't be five-tenths alive.

Pohl, Frederik
The Coming of the Quantum Cats
22 August 1983
4:20 A.M. Senator Dominic DeSota (p. 57)

And first, what is chance? The ancients distinguished between phenomena seemingly obeying harmonious laws, established one and for all, and those which they attributed to chance; these were the ones unpredictable because rebellious to all law. In each domain the precise laws did not decide everything, they only drew limits between which chance might act. In this conception the word chance had a precise and objective meaning: what was chance for one was also chance for another and even for the gods.

Poincaré, Henri
The Foundations of Science
Science and Method (p. 395)

Every phenomenon, however minute, has a cause; and a mind infinitely powerful, infinitely well-informed about the laws of nature, could have foreseen it from the beginning of the centuries. If such a mind existed, we could not play with it at any game of chance; we should always lose.

Poincaré, Henri
The Foundations of Science
Science and Method (p. 395)

Chance is only the measure of our ignorance.

Poincaré, Henri
The Foundations of Science
Science and Method (p. 395)

The greatest bit of chance is the birth of a great man. It is only by chance that the meeting of two germinal cells, of different sex, containing precisely, each on its side, the mysterious elements whose mutual reaction must produce the genius. One will agree that these elements must be rare and that their meeting is still more rare. How slight a thing it would have required to deflect from its route the carrying spermatozoon. It would have suffered to deflect it a tenth of a millimeter and Napoleon would not have been born and the destinies of a continent would have been changed. No example can better make us understand the veritable characteristics of chance.

Poincaré, Henri
The Foundations of Science
Science and Method (pp. 410–1)

All chance, direction, which thou canst not see;
All discord, harmony, not understood,
All partial evil, universal good:
And, spite of Pride, in erring Reason's spite,
One truth is clear, *"Whatever is, is Right."*

Pope, Alexander
The Complete Poetical Works of POPE
An Essay on Man
Epistle I, 289

Wisdom liketh not chance.

Proverb, English

Thus we must content our selves for the most part, to find out how Things may be; without pretending to come to a certain knowledge and determination of what they really are.
[We must for the most part be content with probability.]

Rohault, Jacques
Rohault's System of Natural Philosophy
Volume I, Part I, Chapter 3, 3

I long ago came to the conclusion that all life is 6 to 5 against.

Runyon, Damon
Collier's
A Nice Place
8 September 1934 (p. 8)

There's no such thing as chance;
And what to us seems merest accident
Springs from the deepest source of destiny.

Schiller, Friedrich
Early Dramas
The Death of Wallenstein
Act II, Scene III

Consider that chance, which, with error, its brother, and folly, its aunt, and malice, its grandmother, rules in this world; which every year and every day, by blows great and small, embitters the life of every son of earth, and yours too.

Schopenhauer, Arthur
Parerga and Paralipomena: Short Philosophical Essays
Wisdom of Life: Aphorisms

Chance will not do the work—Chance sends the breeze;
But if the pilot slumbers at the helm,
The very wind that wafts us toward the port
May dash us on the Shelves—The steersman's part is vigilance,
Blow it or rough or smooth.

Scott, Sir Walter
Fortunes of Nigel
Chapter XXII

Give up yourself merely to chance and hazard,
From firm security.

Shakespeare, William
The Complete Works of William Shakespeare
Anthony and Cleopatra
Act III, Scene 7, l. 48

As things but done by chance.

Shakespeare, William
The Complete Works of William Shakespeare
Anthony and Cleopatra
Act V, Scene 2, l. 120

Wherein I spake of most disastrous chances . . .

Shakespeare, William
The Complete Works of William Shakespeare
Othello, The Moor of Venice
Act I, Scene 3, l. 134

Portia. In terms of choice I am not solely led
By nice direction of a maiden's eyes;
Besides, the lottery of my destiny
Bars me the right of voluntary choosing.

Shakespeare, William
The Complete Works of William Shakespeare
The Merchant of Venice
Act II, Scene 1, l. 13

Portia. You must take your chance,
And either not attempt to choose at all
Or swear before you choose, if you choose wrong . . .

Shakespeare, William
The Complete Works of William Shakespeare
The Merchant of Venice
Act II, Scene 1, l. 38

Come, bring me unto my chance.

Shakespeare, William
The Complete Works of William Shakespeare
The Merchant of Venice
Act II, Scene 1, l. 43

If chance will have me King, why, chance may crown me . . .

Shakespeare, William
The Complete Works of William Shakespeare
Macbeth
Act I, Scene 3, l. 143

Florizel . . .
But as the unthought-on accident is guilty
To what we wildly do, so we profess
Ourselves to be the slaves of chance and flies
Of every wind that blows.

Shakespeare, William
The Complete Works of William Shakespeare
The Winter's Tale
Act IV, Scene 4, l. 548

Of Fate, and Chance, and God, and Chaos old . . .

Shelley, Percy Bysshe
The Poems of Percy Bysshe Shelley
Prometheus Unbound
Act II, Scene III, l. 92

Fate, Time, Occasion, Chance and Change—to these all things are subject.

Shelley, Percy Bysshe
The Poems of Percy Bysshe Shelley
Prometheus Unbound
Act II, Scene IV, l. 119

And grasps the skirt of happy chance . . .

Tennyson, Alfred Lord
The Poems and Plays of Tennyson
In Memoriam A.H.H.
Part 1, xiv

Blessed be the gods, by whose aid things happen that we wouldn't even dare hope for!

Terence
Phormio
Act V, Scene 4, l. 757
Quoted in George E. Duckworth's
The Complete Roman Drama

So they cast lots, and the lot fell upon Jonah.

The Bible
Jonah 1:7

. . . the race is not to the swift, nor the battle to the strong . . .

The Bible
Proverbs 16:33

... chance is an empty word without sense, but which is always opposed to that of intelligence, without attaching any determinate, or any certain idea.

Thiery, Paul Henri, Baron d'Holbach
The System of Nature
Volume I
Chapter 5 (p. 71)

For sometimes the course of things is as arbitrary as the plans of man; indeed this is why we usually blame chance for whatever does not happen as we expected.

Thucydides
The History of the Peloponnesian War
I, 140

Why did it happen in this and not in some other way? Because it happened so! "*Chance* created the situation; *genius* utilized it," says history.

But what is *chance*? What is *genius*?

The words *chance* and *genius* do not denote any really existing thing and therefore cannot be defined.

Tolstoy, Leo
War and Peace
First Epilogue, Chapter II

No more chance than a snowball in Hell.

Unknown

Omnium versatur urna serius ocius sors exitura.
[Age at death is a chance variable.]

Unknown

Since Fortune sways to the world . . .
[Chance sways all.]

Virgil
Quoted in James Lonsdale's
The Works of Virgil
The Eclogues
IX, l. 5

COMMON SENSE

Common sense is not really so common.

Arnauld, Antoine
The Art of Thinking: Port-Royal Logic
First Discourse (p. 9)

The double analysis kills the single analysis, and the treble kills the double, until at last a sufficiency of statistics comes very near to common sense.

Belloc, Hilaire
The Silence of the Sea
On Statistics (p. 173)

And then he knew that something within him more powerful than his common-sense would force him to stake that five-franc piece. He glanced furtively at the crowd to see whether anyone was observing him. No. Well, it having been decided to bet, the next question was, how to bet? Now, Henry had read a magazine article concerning the tables at Monte Carlo, and, being of a mathematical turn, had clearly grasped the principles of the game. He said to himself, with his characteristic caution: "I'll wait till red wins four times running, and then I'll stake on the black."

("But surely," remarked the logical superior person in him, "You don't mean to argue that a spin of the ball is affected by the spins that have proceeded it? You don't mean to argue that because red wins four times, or fifty times, running, black is any the more likely to win at the next spin?" "You shut up!" retorted the human side of him crossly. "I know all about that.")

Bennett, Arnold
A Great Man
Chapter XXV (pp. 245–6)

Statistics are no substitute for common sense.

Bialac, Richard N.
Quoted in Paul Dickson's
The Official Explanations (p. B-14)

There is no more remarkable feature in the mathematical theory of probability than the manner in which it has been found to harmonize with, and justify, the conclusions to which mankind have been led, not by reasoning, but by instinct and experience, both of the individual and of the race. At the same time it has corrected, extended, and invested them with a definiteness and precision of which these crude, though sound, appreciations of common sense were till then devoid.

Crofton, M.W.
Encyclopaedia Britannica
9th Edition
Probability

. . . common sense is nothing more than a deposit of prejudices laid down in the mind before you reach eighteen.

Einstein, Albert
Quoted in Eric T. Bell's
Mathematics: Queen and Servant of Science (p. 42)

What is common sense? That which attracts the least opposition: that which brings most agreeable and worthy results.

Howe, E.W.
Sinner Sermons (p. 7)

We know that the probability of a well-established induction is great, but, when we are asked to name its degree, we cannot. Common sense tells us that some inductive arguments are stronger than others, and that some are very strong. But how much stronger or how strong we cannot express.

Keynes, John Maynard
A Treatise on Probability
Chapter XXI (p. 259)

One sees in this essay that the theory of probabilities is basically only common sense reduced to a calculus.

Laplace, Pierre-Simon
A Philosophical Essay on Probabilities (p. 124)

CORRELATION

There is no correlation between the cause and the effect. The events reveal only an aleatory determination, connected not so much with the imperfection of our knowledge as with the structure of the human world.

Aron, Raymond
The Opium of the Intellectuals
Chapter VI (p. 163)

"You know those penetration figures?"

"Mm."

"Well, there's a positive correlation between penetration and the height of the man firing."

"Easy," I said. "The taller the man, the more rarefied the atmosphere and the less the air resistance."

Balchin, Nigel
The Small Back Room (p. 8)

"Very true," said the Duchess: "flamingos and mustard both bite. And the moral of that is 'Birds of a feather flock together.'"

"Only mustard isn't a bird." Alice remarked.

"Right as usual," said the Duchess: "what a clear way you have of putting things!"

Carroll, Lewis
The Complete Works of Lewis Carroll
The Mock Turtle's Story

Reading the twenty-sixth chart, one correlation suddenly occurred to Jason. Although the patients did not share physical symptoms, their charts showed a predominance of high-risk social habits. They were overweight, smoked heavily, used drugs, drank too much, and failed to exercise, or combined any and all of these unhealthy practices; they were men and women who were eventually destined to have severe medical problems. The shaking fact was that they deteriorated so quickly. And why the sudden upswing in deaths. People weren't indulging in vices more than they were a year ago. Maybe it was a kind of statistical equalizing. They'd been lucky and now the numbers were catching up to them.

Cook, Robin
Mortal Fear
Chapter 11 (p. 220)

The well-known virtue of the experimental method is that it brings situational variables under tight control. It thus permits rigorous tests of hypotheses and confidential statements about causation. The correlational method, for its part, can study what man has not learned to control. Nature has been experimenting since the beginning of time, with a boldness and complexity far beyond the resources of science. The correlator's mission is to observe and organize the data of nature's experiments.

Cronbach, L.J.
The American Psychologist
The Two Disciplines of Scientific Psychology
Volume 12, November 1957 (p. 672)

Hall's Law: There is a statistical correlation between the number of initials in an Englishman's name and his social class (the upper class having significantly more than three names, while members of the lower class average 2.6).

Dickson, Paul
The Official Rules (p. H-80)

The futile elaboration of innumerable measures of correlation, and the evasion of the real difficulties of sampling problems under cover of a contempt for small samples, were obviously beginning to make its pretensions ridiculous. These procedures were not only ill-aimed, but for all their elaboration, not sufficiently accurate.

Fisher, Sir Ronald A.
Statistical Methods for Research Workers (p. v)

"Co-relation or correlation of structure" is a phrase much used in biology, and not least in that branch of it which refers to heredity, and the idea is even more frequently present than the phrase but I am not aware of any previous attempt to define it clearly, to trace its mode of action in detail, or to show how it measures its degree.

Galton, Francis
Proceedings of the Royal Society of London
Co-relations and Their Measurements, Chiefly for Anthropometric Data
Volume 45, 1888

It had appeared from observation, and it was fully confirmed by this theory, that such a thing existed as an "Index of Correlation", that is to say, a fraction, now commonly written r, that connects with close approximation every value of the deviation on the part of the subject, with the *average* of all the associated deviations of the Relative . . .

Galton, Francis
Memories of My Life
Chapter XX

It is now beginning to be generally understood, even by merely practical statisticians, that there is truth in the theory that all variability is much the same kind.

Galton, Francis
North American Review
Kinship and Correlation
Volume 150, Part II, April 1890 (pp. 427–8)

I can only say that there is a vast field of topics that fall under the laws of correlation, which lies quite open to the research of any competent person who cares to investigate it.

Galton, Francis
North American Review
Kinship and Correlation
Volume 150, Part II, April 1890 (p. 431)

Biological phenomena in their numerous phases, economic and social, were seen to be only differentiated from the physical by the intensity of their correlations. The idea Galton placed before himself was to represent by a single quantity the degree of relationships, or of partial causality between the different variables of our everchanging universe.

Pearson, Karl
The Life, Letters, and Labours of Francis Galton
Volume IIIA, Chapter XIV (p. 2)

The quantity of the correlation is inversely proportional to the density of the control (the fewer the facts, the smoother the curves).

Unknown

DATA

There is no substitute for honest, thorough, scientific effort to get correct data (no matter how much of it clashes with preconceived ideas). There is no substitute for actually reaching a correct claim of reasoning. Poor data and good reasoning give poor results. Good data and poor reasoning give poor results. Poor data and poor reasoning give rotten results.

Berkeley, Edmund C.
Computers and Automation
Right Answers—A Short Guide for Obtaining Them
Volume 18, Number 10, September 1969 (p. 20)

Lots of people bring you false information.

Berkeley, Edmund C.
Computers and Automation
Right Answers—A Short Guide for Obtaining Them
Volume 18, Number 10, September 1969 (p. 20)

Anyone can easily misuse good data.

Deming, William Edwards
Some Theory of Sampling (p. 18)

There is only one kind of whiskey, but two broad classes of data, good and bad.

Deming, William Edwards
The American Statistician
On the Classification of Statistics
Volume 2, Number 2, April 1948 (p. 16)

Scientific data are not taken for museum purposes; they are taken as a basis for doing something. If nothing is to be done with the data, then there is no use in collecting any. The ultimate purpose of taking data is to provide a basis for action or a recommendation for action. The step intermediate between the collection of data and the action is prediction.

Deming, William Edwards
Journal of the American Statistical Association
On a Classification of the Problems of Statistical Inference
Volume 37, Number 218, June 1942 (p. 173)

Data are often presented in a form that is not immediately clear. The reader can then either ignore the data, analyze them himself, or return them to the author for *him* to analyze.

Ehrenberg, A.S.C.
Data Reduction (p. 1)

It does not follow that because something *can* be counted it therefore *should* be counted.

Enarson, Harold L.
Speech to Society for College and University Planning, September 1975

No human mind is capable of grasping in its entirety the meaning of any considerable quantity of numerical data.

Fisher, Sir Ronald A.
Statistical Methods for Research Workers (p. 6)

I can only suggest that, as we are practically without data, we should endeavor to obtain some.

Freeman, R. Austin
A Certain Dr. Thorndyke
Thorndyke Takes up the Inquiry

My data were very lax but this method of treatment got all the good out of them that they possessed.

Galton, Francis
Natural Inheritance
Schemes of Distribution and of Frequency (p. 48)

Still, it is an error to argue in front of your data. You find yourself insensibly twisting them around to fit your theories.

Holmes, Sherlock
in Arthur Conan Doyle's
The Complete Sherlock Holmes
The Adventure of Wisteria Lodge

"Data! Data! Data!" he cried impatiently. "I can't make bricks without clay."

Holmes, Sherlock
in Arthur Conan Doyle's
The Complete Sherlock Holmes
The Adventure of the Copper Beeches

It is a capital mistake to theorize before one has the data.

Holmes, Sherlock
in Arthur Conan Doyle's
The Complete Sherlock Holmes
Scandal in Bohemia

"No data yet," he answered. "It is a capital mistake to theorize before you have all of the evidence. It biases the judgment."

Holmes, Sherlock
in Arthur Conan Doyle's
The Complete Sherlock Holmes
A Study in Scarlet

If you can't have an experiment, do the best you can with whatever data you can gather, but do be very skeptical of historical data and subject them to all the logical tests you can think of.

Hooke, Robert
Quoted in J.M. Tanur's
Statistics: A Guide to the Unknown
Statistics, Sports, and Some Other Things

To the optical astronomer, radio data serves like a good dog on a hunt.

Hoyle, Fred
Galaxies, Nuclei and Quasars (p. 43)

By no process of sound reasoning can a conclusion drawn from limited data have more than a limited application.

Mellor, J.W.
Higher Mathematics for Students of Chemistry and Physics (p. 4)

When a man of science speaks of his "data", he knows very well in practice what he means. Certain experiments have been conducted, and have yielded certain observed results, which have been recorded. But when we try to define a "datum" theoretically, the task is not altogether easy. A datum, obviously, must be a fact known by perception. But it is very difficult to arrive at a fact in which there is no element of inference, and yet it would seem improper to call something a "datum" if it involved inferences as well as observation. This constitutes a problem . . .

Russell, Bertrand A.
The Analysis of Matter
Chapter XIX (p. 187)

The individual source of the statistics may easily be the weakest link. Harold Cox tells a story of his life as a young man in India. He quoted some statistics to a Judge, an Englishman, and a very good fellow. His friend said, "Cox, when you are a bit older, you will not quote Indian statistics with that assurance. The Government are very keen on amassing statistics—they collect them, and they raise them to the nth power, take the cube root and prepare wonderful diagrams. But what you must never forget is that every one of those figures comes in the first instance from the *chowty dar* (village watchman), who just puts down what he damn pleases."

Stamp, Josiah
Some Economic Factors in Modern Life
Chapter VII (p. 258)

We have no scientific data whatever on clock-eating and hence no controlled observation.

Thurber, James
Lanterns and Lances
The Last Clock

In general, it is necessary to have some data on which to calculate probabilities . . . Statisticians do not evolve probabilities out of their inner consciousness, they merely calculate them.

Tippett, L.C.
The World of Mathematics
Sampling and the Standard Error
Volume 3 (p. 1486)

Sint ut sunt aut non sint.
[Accept them as they are or deny their existence.]

Unknown

If at first you don't succeed, transform your data set.

Unknown

DEFINITIONS

An observation with an abnormally large residual will be referred to as an *outlier*. Other terms in English are "wild", "straggler", "sport" and "maverick"; one may also speak of a "discordant", "anomalous" or "aberrant" observation.

Anscombe, F.J.
Technometrics
Rejection of Outliers
Volume 2, 1960

Common knowledge is, in fact, nothing else than the raw material which, assorted, refined and chemically transmuted, has served as the basic substance of its vastly elaborated synthesis.

Barry, Frederick
The Scientific Habit of Thought (p. 20)

Die, *n.* The singular of "dice". We seldom hear the word, because there is a prohibitory proverb, "Never say die". At long intervals, however, some one says: "The die is cast", which is not true, for it is cut. The word is found in an immortal couplet by that eminent poet and domestic economist, Senator Depew:

A cube of cheese no larger than a die
May bait the trap to catch a nibbling mie.

Bierce, Ambrose
The Devil's Dictionary

Faith, *n.* Belief without evidence in what is told by one who speaks without knowledge, of things without parallel.

Bierce, Ambrose
The Devil's Dictionary

Reason, *v.i.* To weigh probabilities in the scales of desire.

Bierce, Ambrose
The Devil's Dictionary

Indecision, *n.* The chief element of success; "for whereas", said Sir Thomas Brewbold, "there is but one way to do nothing and diverse ways to do something, whereof, to a surety, only one is the right way, it followeth that he who from indecision standeth still hath not so many chances of going astray as he who pusheth forwards"—a most clear and satisfactory exposition of the matter.

"Your prompt decision to attack", said General Grant on a certain occasion to General Gordon Granger, "was admirable; you had but five minutes to make up your mind in."

"Yes, Sir," answered the victorious subordinate, "it is a great thing to know exactly what to do in an emergency. When in doubt whether to attack or retreat I never hesitate a moment—I toss up a copper."

"Do you mean to say that's what you did this time?"

"Yes, General; but for Heaven's sake don't reprimand me: I disobeyed the coin."

Bierce, Ambrose
The Devil's Dictionary

A chunk is a convenient slice of a population.

Deming, William Edwards
Some Theory of Sampling (p. 14)

assessed probability: One manipulated by the Internal Revenue Service.

assignable cause: The cause that takes the rap when the process runs amok.

best estimate: In the theory of estimation, an estimate having optimum qualities under conditions almost never met in practice.

commode: Term applied to each mode of a bimodal distribution.

data: 1. Brandname for the products from down here. 2. Plural of datum, meaning reference point. When there are more than one, they *almost always* conflict. 3. Deified numbers.

expected value: One that the sample average will almost never equal.

posterior probability: A result arrived at by the application of an elegant mathematical formula to nothing more than seat-of-the-pants reasoning.

probability: An erudite measure of ignorance. Being dimensionless, it is best used with a dimensional measure, especially a grain of salt.

random normal deviate: A contradiction in terms, since deviates are abnormal.

regression fallacy: The naive belief that regression analysis is a cure-all. Those who entertain it are known as regressions, and their way is hard. They regress first and think afterward.

scatterbrain: 1. A Bayesian whose beliefs have been randomized in order to facilitate deriving *personal probabilities* without systematic bias. 2. A Classicist who scatters test levels to the wind, hoping that one will prove significant.

sequential analysis: A systematic procedure for generating second guesses.

statistics: 1. A form of lying that is neither black, white, nor color. 2. An attempt to analyze *data*—rare and archaic. 3. A disorderly, but not quite random, progress from datum to datum.

Durand, David
The American Statistician
A Dictionary for Statismagicians
Volume 24, Number 3, June 1970 (p. 21)

When there is no explanation, they give it a name, which immediately explains everything.

Fabing, Harold
Mar, Ray
Fischerisms (p. 4)

Thinking in words, consciousness is behavior, experiment is measurement.

Green, Celia
The Decline and Fall of Science
Aphorisms (p. 172)

Misinforming people by use of statistical material might be called statistical manipulation; in a word (though not a very good one), statisticulation.

Huff, Darrell
How to Lie with Statistics (p. 100)

summation convention *n.* A mathematicians' shindig held each year in the Kronecker Delta.

Kelly-Bootle, Stan
The Devil's DP Dictionary

standard deviation *n.* A sexual activity formerly considered perverted but now universally practiced and accepted.

A DP Freudian writes: "I divide my patients into two broad categories: those who are turned on by normally distributed curves and those who are not. Do not fret, I tell them all. One person's meat is another person's Poisson. That soon gets the idiots off my couch, out of my sample, and into my accounts payable . . ."

Kelly-Bootle, Stan
The Devil's DP Dictionary

map *n*. The imponderable correspondence between two sets, one of which is unknown (called the domain), while the other (the *range*) is unknowable.

Kelly-Bootle, Stan
The Devil's DP Dictionary

The words figure and fictitious both derive from the same Latin root, fingere. Beware!

Moroney, M.J.
Facts from Figures
Scatter (p. 56)

Innumeracy, an inability to deal comfortably with the fundamental notions of numbers and chance, plagues far too many otherwise knowledgeable citizens.

Paulos, John Allen
Innumeracy (p. 3)

Inference, *n*. A mysterious process allowing us to reach a conclusion that is desired.

An old sea captain kept a personal diary. On his sixty-fifth birthday he wrote: "Awoke this morning with a fine erection, couldn't bend it with both hands." On his seventieth birthday he wrote: "Awoke this morning with a fine erection; couldn't bend it with both hands." On his seventy-fifth birthday he wrote: "Awoke this morning with a fine erection; could barely bend it with both hands. Must be getting stronger."

Plonk, Phineas
Quoted in Edmund H. Volkart's
The Angel's Dictionary

A posit is a statement which we treat as true although we do not know whether it is so.

Reichenbach, Hans
The Rise of Scientific Philosophy (p. 240)

Statistics, *n. pl.* The collection, analysis, and interpretation of numerical data in such a way as to be understood by computers and misunderstood by everyone else.

Volkhart, Edmund H.
The Angel's Dictionary

Define your terms, you will permit me again to say, or we shall never understand one another.

Voltaire
The Portable Voltaire
Philosophical Dictionary
Miscellany (p. 225)

A precise and universally acceptable definition of the term 'nonparametric' is not presently available.

Walsh, John E.
Handbook of NonParametric Statistics
Volume 1, Chapter 1 (p. 2)

It must always be remembered that man's body is what it is through having been molded into its present shape by the chances and changes of an immense time . . .

Samuel Butler –

(See p. 36)

DEGREES OF FREEDOM

Degrees of freedom. The number of fetters on the statistician. The number of d.f. is usually considered self-evident—except for the analysis of data that have not appeared in a textbook.

Durand, David
The American Statistician
A Dictionary for Statismagicians
Volume 24, Number 3, June 1970 (p. 21)

The conception of degrees of freedom is not altogether easy to attain . . .

Tippett, L.C.
The Method of Statistics (p. 64)

DESIGN OF EXPERIMENTS

"The first thing I've got to do" said Alice to herself as she wandered in the woods, "is to grow to my right size again; and the second thing is to find my way into that lovely garden. I think that will be the best plan." It sounded an excellent plan, no doubt; the only difficulty was that she had not the smallest idea how to set about it; . . .

Carroll, Lewis
The Complete Works of Lewis Carroll
The Rabbit Sends in a Little Bill

A lady declares that by tasting a cup of tea made with milk she can discriminate whether the milk or the tea infusion was first added to the cup. We will consider the problem of designating an experiment by means of which this assertion can be tested.

Fisher, Sir Ronald A.
The Design of Experiments (p. 13)

If you're trying to establish cause-and-effect relationships, do try to do so with a properly designed experiment.

Hooke, Robert
Quoted in J.M. Tanur's
Statistics: A Guide to the Unknown
Statistics, Sports, and Some Other Things

One day when I was a junior medical student, a very important Boston surgeon visited the school and delivered a great treatise on a large number of patients who had undergone successful operations for vascular reconstructions. At the end of the lecture, a young student at the back of the room timidly asked, "Do you have any controls?" Well, the great surgeon drew himself up to his full height, hit the desk, and said, "Do you mean did I not operate on half the patients?" The hall grew very quiet then. The voice at the back of the room very hesitantly replied, "Yes, that's what I had in mind." Then the visitor's fist really came down as he thundered, "Of course not. That would have doomed half of them to their death." It was absolutely silent then, and one could scarcely hear the small voice ask, "Which half?"

Peacock, E.E.
Medical World News
September 1, 1972 (p. 45)

A mighty maze! but not without a plan . . .

Pope, Alexander
The Complete Poetical Works of POPE
An Essay on Man, Epistle I, l. 6

A committee or an investigator considering a scheme of experiments should first . . . ask whether each experiment or question is framed in such a way that a definite answer can be given. The chief requirement is simplicity; only one question should be asked at a time.

Russell, E.J.
Journal of the Ministry of Agriculture of Great Britain
Field Experiments: How They are Made and What They Are
Volume 32, 1926 (p. 989)

For which of you intending to build a tower, sitteth not down first, and counteth the cost, whether he have sufficient to finish it?

The Bible
Luke 14:28

DICE

... to repeat the same throw ten thousand times with the dice would be impossible, whereas to make it once or twice is comparatively easy.

Aristotle
On the Heavens
Book II, Chapter XII

Appeal: *v.t.* In law, to put the dice into the box for another throw.

Bierce, Ambrose
The Devil's Dictionary

Four dice are cast and a Venus throw results—that is chance; but do you think it would be chance, too, if in one hundred casts you made one hundred Venus throws? It is possible for paints flung at random on a canvas to form the outline of a face; but do you imagine that an accidental scattering of pigments could produce the beautiful portrait of Venus of Cos? Suppose that a hog should form a letter 'A' on the ground with its snout; is that a reason for believing that it could write out Ennius's poem *The Andromache*?

Cicero
Cicero: De Senectute, De Amicitia, De Divinatione
De Divinatione
I. xiii

'Tis fate that flings the dice,
and as she flings
of Kings makes peasants,
and of peasants Kings.

Dryden, John
Works
Volume XV, 1821 Edition (p. 103)

Quantum mechanics is certainly imposing. But an inner voice tells me that it is not yet the real thing. The theory says a lot but does not bring us any closer to the secret of the Old One. I, at any rate, am convinced that He does not throw dice.

Einstein, Albert
Quoted in Ronald W. Clark's
Einstein: The Life and Times (p. 340)

I can, if the worst comes to worst, still realize that God may have created a world in which there are no natural laws. In short, a chaos. But that there should be statistical laws with definite solutions, i.e., laws which compel God to throw the dice in each individual case, I find highly disagreeable.

Einstein, Albert
Quoted in Ronald W. Clark's
Einstein: The Life and Times (p. 340)

Acorns may be food for hogs or rise into magnificent oaks, as the dice of chance decree.

Eldridge, Paul
Maxims for a Modern Man
1849

The first steps in Agriculture, Astronomy, Zoology (those first steps which the farmer, the hunter, and the sailor take), teach that Nature's dice are always loaded; that in her heaps of rubbish are concealed sure and useful results.

Emerson, Ralph Waldo
Nature
Discipline (p. 38)

The dice of God are always loaded.

Emerson, Ralph Waldo
Essays
First Series
Compensation

It therefore seems that Einstein was doubly wrong when he said that God does not play dice. Consideration of particle emission from black holes suggests that God not only plays with dice but that he sometimes throws them where they cannot be seen.

Hawking, S.
Nature
The Breakdown of Physics
Volume 257, 1975 (p. 362)

For dice will run the contrary way
As well is known to all who play . . .

Hood, Thomas
Miss Kilmansegg and Her Previous Leg
Her Misery
l. 2150

They need only adapt to the circumstances that old Lydian tradition
which says that games of chance were invented during great famine.
Men permitted themselves to eat only every second day, and tried to
forget their hunger by playing at draughts and dice.

Lang, Andrew
Lost Leaders
Winter Sports

Un Coup de dés jamis n'abolira le hasard.
[A throw of the dice will never abolish chance.]

Mallarmé, Stéphane
Title of poem in *Poems* (p. 159)

Jacta alea est.
[The die is cast.]

Plutarch
Plutarch's Lives
Caesar

One day in Naples the reverend Galiana saw a man from the Basilicata
who, shaking three dice in a cup, wagered to throw three sixes; and, in
fact, he got three sixes right away. Such luck is possible, you say. Yet the
man succeeded a second time, and the bet was repeated. He put back the
dice in the cup, three, four, five times, and each time he produced three
sixes. 'Sangue di Bacco,' exclaimed the reverend, 'the dice are loaded!'
And they were.

Polya, G.
Patterns of Plausible Inference (p. 74)

I hear the clack—
who cast the dice
on the bathroom tiles?

Ritsos, Yannis
Erotica
Small Suite in Red Major

And by the hazard of the spotted die
Let die the spotted.

Shakespeare, William
The Complete Works of William Shakespeare
Timon of Athens
Act V, Scene 4, l. 34

King Richard. A horse! a horse! my kingdom for a horse!

Catesby. Withdraw, my lord; I'll help you to a horse.

King Richard. Slave, I have set my life upon a cast
And I will stand the hazard of the die:
I think there be six Richmonds in the field;
Five have I slain to-day instead of him.
A horse! a horse! my kingdom for a horse!

Shakespeare, William
The Complete Works of William Shakespeare
The Tragedy of King Richard the Third
Act V, Scene 4, l. 7

Midas in tesseris consultor optimus.
[Midas on the dice gives the best advice.]

Suidas
Collected Works of Erasmus
Adages II vii 1 to III iii 100 (p. 124)

We were shaken into existence, like dice from a box.

Wilder, Thornton
The Eighth Day
II, Illinois to Chile (p. 107)

I of dice possess the science,
And in numbers thus am skilled.

Williams, Monier
The Story of Nala
Book XX (p. 133)

DISTRIBUTIONS

A sea-fight must either take place to-morrow or not, but it is not necessary that it should take place to-morrow, neither is it necessary that it should not take place, yet it is necessary that it either should or should not take place to-morrow.

Aristotle
On Interpretation
Chapter IX

Yet there are Writers, of a Class indeed very different from that of *James Bernoulli*, who insinuate as if the *Doctrine of Probabilities* could have no place in any serious Enquiry; and that Studies of this kind, trivial and easy as they be, rather disqualify a man for reasoning on every other subject. Let the Reader chuse.

de Moivre, Abraham
The Doctrine of Chances (p. 254)

The primary objects of the Gaussian Law of Error were exactly opposed, in one sense, to those to which I applied them. They were to get rid of, or to provide a just allowance for errors. But these errors or deviations were the very thing I wanted to preserve and to know about.

Galton, Francis
Memories of My Life
Chapter XX

It has been objected . . . that I pushed the application of the Law of Frequency of Error somewhat too far. I may have done so . . . ; but I am sure that, with the evidence before me, the applicability of that law is more than justified within . . . reasonable limits.

Galton, Francis
Natural Inheritance
Schemes of Distribution and of Frequency (p. 44)

71

Normality is a myth; there never has, and never will be, a normal distribution.

Geary, R.C.
Biometrika
Testing for Normality
Volume 34, 1947 (p. 241)

If the prior distribution, at which I am frankly guessing, has little or no effect on the result, then why bother; and if it has a large effect, then since I do not know what I am doing how would I dare act on the conclusions drawn?

Hamming, Richard W.
The Art of Probability for Scientists and Engineers (p. 298)

Which Bernoulli do you wish to see—'Hydrodynamics' Bernoulli, 'Calculus' Bernoulli. 'Geodesic' Bernoulli. 'Large Numbers' Bernoulli or 'Probability' Bernoulli?

Harris, Sidney
What's So Funny about Science
Caption to Cartoon

. . . to quote a statement of Poincaré, who said (partly in jest no doubt) that there must be something mysterious about the normal law since mathematicians think it is a law of nature whereas physicists are convinced that it is a mathematical theorem.

Kac, Mark
Statistical Independence in Probability Analysis and Number Theory
Chapter 3, The Normal Law (p. 52)

A mathematician in Reno,
Overcome by the heat and the vino,
Became quite unroulli
Expounding Bernoulli,
And was killed by the crows playing Keno.

Kelly-Bootle, Stan
The Devil's DP Dictionary

A misunderstanding of Bernoulli's theorem is responsible for one of the commonest fallacies in the estimation of probabilities, the fallacy of the maturity of chances. When a coin has come down heads twice in succession, gamblers sometimes say that it is more likely to come down tails next time because 'by the law of averages' (whatever that may mean) the proportion of tails must be brought right some time.

Kneale, W.
Probability and Induction (p. 140)

It has become increasingly apparent over a period of several years that psychologists, taken in the aggregate, employ the chi-square test incorrectly.

Lewis, Don
Burke, C.J.
Psychological Bulletin
The Use and Misuse of the Chi-Square Test
Volume 46, Number 6, November 1949 (p. 433)

Distribute dissatisfaction uniformly.

Lidberg, A.A.
Quoted in Paul Dickson's
The Official Explanations (p. B-21)

Les Expérimentateurs s'imaginent que c'est un théorème de mathématique, et les mathématiciens d'étreun fait expérimental!
[Everybody believes in the [normal approximation], the experimenters because they think it is a mathematical theorem, the mathematicians because they think it is an experimental fact!]

Lippmann, G.
Quoted in D'Arcy Thompson's
On Growth and Form
Volume I (p. 121)

I would therefore urge that people be introduced to [the logistic equation] early in their mathematical education. This equation can be studied phenomenologically by iterating it on a calculator, or even by hand. Its study does not involve as much conceptual sophistication as does elementary calculus. Such study would greatly enrich the student's intuition about nonlinear systems.

Not only in research but also in the everyday world of politics and economics, we would all be better off if more people realized that simple nonlinear systems do not necessarily possess simple dynamical properties.

May, Robert M.
Nature
Simple Mathematical Models with very Complicated Dynamics
Volume 261, June 10, 1976 (p. 467)

We know not to what are due the accidental errors, and precisely because we do not know, we are aware they obey the law of Gauss. Such is the paradox.

Poincaré, Henri
The Foundations of Science
Science and Method (p. 406)

Roger has tried to explain to her the V-bomb statistics: the difference between distribution . . . She's almost got it; nearly understands his Poisson equation . . .

Pynchon, Thomas
Gravity's Rainbow (p. 54)

But a hardon, that's either there, or it isn't. Binary, elegant. The job of observing it can even be done by a *student*.

Pynchon, Thomas
Gravity's Rainbow (p. 84)

You have two chances—
One of getting the germ
And one of not.
And if you get the germ
You have two chances—
One of getting the disease
And one of not.
And if you get the disease
You have two chances—
One of dying
And one of not.
And if you die—
Well, you still have two chances.

Unknown

When you get an 8 on the midterm, there ain't a curve in the world that can save you.

Unknown

An exterminator made this contribution
On rats arriving in random profusion
"I know nothing of math,
Probability of stats,
But I handle 'em with Poisson distributions."

Unknown
Quoted in Arnold O. Allen's
*Probability, Statistics, and Queueing Theory with
Computer Science Applications* (p. 86)

Keep your hyperexponential away from me!

Unknown
Quoted in Arnold O. Allen's
*Probability, Statistics, and Queueing Theory with
Computer Science Applications* (p. 178)

Monique is exponentially distributed.

Unknown
Quoted in Arnold O. Allen's
*Probability, Statistics, and Queueing Theory with
Computer Science Applications* (p. 178)

Socrates took Poisson.

Unknown
Quoted in Arnold O. Allen's
*Probability, Statistics, and Queueing Theory with
Computer Science Applications* (p. 178)

The
normal
law of error
stands out in the
experience of man-
kind as one of the broad-
est generalizations of natural
philosophy. It serves as the guiding
instrument in researches in the physical
and social sciences and in medicine, agriculture and
engineering. It is an indispensable tool for the analysis and the
interpretation of the basic data obtained by observation and experiment.

Youden, W.J.
Experimentation and Measurement (p. 55)
See also
The American Statistician
April–May 1950 (p. 11)

ERROR

If frequently I fret and fume,
  And absolutely will not smile,
I err in company with Hume,
  Old Socrates and T. Carlyle.

Adams, Franklin
Tobogganing on Parnassus
Erring in Company

One sufficiently erroneous reading can wreck the whole of a statistical analysis, however many observations there are.

Anscombe, F.J.
Technometrics
Rejection of Outliers
Volume 2, 1960 (p. 226)

The problem of error has preoccupied philosophers since the earliest antiquity. According to the subtle remark made by a famous Greek philosopher, the man who makes a mistake is twice ignorant, for he does not know the correct answer, and he does not know that he does not know it.

Borel, Emile
Probability and Certainty
Chapter 9 (p. 114)

For error and mistake are infinite,
But truth has but one way to be i' th' right.

Butler, Samuel
The Poetical Works
Miscellaneous Thoughts
l. 114

An error is simply a failure to adjust immediately from a preconception to an actuality.

Cage, John
Silence 1961
45' for a Speaker

No error at all! They were positively steeped in error!

Carroll, Lewis
The Complete Works of Lewis Carroll
A Tangled Tale

Pepys probably did not much increase his popularity in the Grafton by getting Dartmouth to call for the dead-reckoning from twelve different persons on board, especially as this was done before they sighted land. Their errors were subsequently found to be very considerable—one was as much as seventy leagues out! It is interesting to note that the inference drawn from these discrepancies was that the chart must be wrong, and it was corrected accordingly.

Chappell, Edwin
The Tangier Papers of Samuel Pepys (p. xxxviii)

Mal-information is more hopeless than non-information; for error is always more busy than ignorance.

Colton, Charles Caleb
Lacon: or many things in a few words (p. 2)

Man, on the dubious waves of error toss'd.

Cowper, William
Cowper: Poetical Works
Truth
l. 1

O mathematicians, throw light on this error.

da Vinci, Leonardo
The Notebooks of Leonardo da Vinci
Volume I
Philosophy (p. 64)

If someone made a mistake he would drawl, "Hell that's why they make erasers."

Darrow, Clarence
Quoted in Irving Stone's
Clarence Darrow for the Defense (p. 75)

Precision is expressed by an international standard, viz., the standard error. It measures the average of the difference between a complete coverage and a long series of estimates formed from samples drawn from this complete coverage by a particular procedure or drawing, and processed by a particular estimating formula.

Deming, William Edwards
Journal of the American Statistical Association
On the Presentation of the Results of Sample Surveys as Legal Evidence
Volume 49, Number 268, December 1954 (p. 820)

Errors, like straws, upon the surface flow,
He who would search for pearls must dive below.

Dryden, John
The Poetical Works of Dryden
All for Love, Prologue, l. 25

However we define error, the idea of calculating its extent may appear paradoxical. A science of errors seems a contradiction in terms.

Edgeworth, Francis Ysidro
Journal of the Royal Statistical Society
Volume 53 (p. 462)

Error is Prolific.

Erasmus, Desiderius
Epicureus

No error is harmless.

Evans, Bergen
The Natural History of Nonsense
A Tale of a Tub

The phrase "Errors of the Second Kind", although apparently only a harmless piece of technical jargon, is useful as indicating the type of mental confusion in which it was coined.

Fisher, Sir Ronald A.
Journal of the Royal Statistical Society
Statistical Methods and Scientific Induction
Series B, Number 17, 1955 (p. 73)

It is doubtful if "Student" ever realized the full importance of his contribution to the Theory of Errors. From correspondence with him before the War . . . I should form a confident judgment that at that time certainly he did not see how big a thing he had done . . . Probably he felt, had the problem really been so important as it had once seemed, the leading authorities in English statistics would have at least given him the encouragement of recommending the use of his method; and better still, would have sought to gain similar advantages in more complex problems. Five years, however, passed without the writers in *Biometrika*, the journal in which he had published, showing any sign of appreciating the significance of his work. This weighty apathy must greatly have chilled his enthusiasm . . . It was sixteen years before, in 1928, the system of tests of which Student was the prototype was logically complete. Only during the thirteen years which have since passed has "Student's" work found its proper place as an experiment resource.

Fisher, Sir Ronald A.
Annals of Eugenics
Student
Volume 9, 1939 (p. 5)

No vehement error can exist in this world with impunity.

Froude, James Anthony
Short Studies on Great Subjects
Spinoza

An error? What error?

Gilbert, W.S.
Sullivan, Arthur
The Complete Plays of Gilbert and Sullivan
The Pirates of Penzance
Act I

Nature itself cannot err.

Hobbes, Thomas
Leviathan
Part I, Chapter IV

The greatest follies are often composed, like the largest ropes, or a multitude of strands.

Hugo, Victor
Les Misérables
Cosette
Book V, Chapter 10

It sounds paradoxical to say the attainment of scientific truth has been effected, to a great extent, by the help of scientific errors.

Huxley, Thomas H.
Method and Results
The Progress of Science (p. 63)

There is no greater mistake than the hasty conclusion that opinions are worthless because they are badly argued.

Huxley, Thomas H.
Method and Results
Natural Rights and Political Rights (p. 369)

... irrationally held truths may be more harmful than reasoned errors.

Huxley, Thomas H.
Collected Essays
The Coming of Age of "The Origin of Species"
Volume II

... quantities which are called *errors* in one case, may really be most important and interesting phenomena in another investigation. When we speak of eliminating error we really mean disentangling the complicated phenomena of nature.

Jevons, W.S.
The Principles of Science
Chapter 15 (p. 339)

... When I make a mistake, it's a beaut.

Manners, William
Patience and Fortitude (p. 219)

... the errors are not the art, but in the artifiers.

Newton, Sir Isaac
Mathematical Principles of Natural Philosophy
Preface to the First Edition

In those sciences of measurement which are the least subject to error— meteorology, geodesy, and metrical astronomy—no man of self-respect ever now states his results, without affixing to it its *probable error*; and if this practice is not followed in other sciences it is because in those the probable errors are too vast to be estimated.

Peirce, Charles Sanders
Philosophical Writing of Peirce (p. 3)

A final word about the theory of errors. Here it is that the causes are complex and multitude. To how many snares is not the observer exposed, even with the best instruments.

Poincaré, Henri
The Foundations of Science
Science and Method (p. 402)

The best may slip, and the most cautious fall;
He's more than mortal that ne'er err'd at all.

Pomfret, John
The Poetical Works of John Pomfret
Love Triumphant over Reason
l. 145

I will stand on, and continue to use, the figures I have used, because I believe they are correct. Now, I'm not going to deny that you don't now and then slip up on something; no one bats a thousand.

Reagan, Ronald
Washington Post
On Bandwagon, Reagan Seeks to Stiffen Credibility Grip
20 April 1980 (A8)

Always expect to find at least one error when you proofread your own statistics. If you don't, you are probably making the same mistake twice.

Russell, Cheryl
Quoted in Tom Parker's
Rules of Thumb (p. 124)

One cannot too soon forget his errors . . .

Thoreau, Henry David
Winter
9 Jan 1842

For the Bureau has worked hard to learn the accuracy of its measurements and it supplies with each weight a certificate indicating how much the weight may differ from exactly one pound. The calibration of the weight is valuable *just because* its possible error is known. When the Bureau of the Census makes an enumeration, there are errors, which they acknowledge. They know the extent of the errors from many sources and they try to learn more about those from others . . . It is far easier to put out a figure, than to accompany the figure with a wise and reasoned account of its liability to systematic and fluctuating errors. Yet if the figure is . . . to serve as the basis of an important decision, the accompanying amount may be more important than the figures themselves.

Tukey, John W.
The American Statistician
Memorandum on Statistics in the Federal Government
Volume 3, Number 5, February 1949 (p. 9)

A Type III error is a good solution to the wrong problem.

Unknown

A Type IV error is a wrong solution to the wrong problem.

Unknown

A standard error is just as bad as any other error.

Watson, Alfred N.
Statement made at a meeting of the American Statistical
Association, Chicago, 1942

There is great room for error here.

Whitehead, Alfred North
Science and the Modern World
Chapter II

The dice of God are always loaded.
Ralph Waldo Emerson – (See p. 68)

EXPERIMENT

But the method of experiment which men now make use of is blind and stupid: and so, while they wander and stray without any certain way, but only take counsel from the occurrence of circumstances, they are carried about to many points, but advance little; . . .

Bacon, Francis
The Novum Organon
First Book, 70

If an experiment works, something has gone wrong.

Bloch, Arthur
Murphy's Law
Finangle's First Law (p. 15)

The experiment may be considered a success if no more than 50% of the observed measurements must be discarded to obtain a correspondence with the theory.

Bloch, Arthur
Murphy's Law
Maier's Law: Corollary (p. 47)

Just an experiment first, for candour's sake.

Browning, Robert
The Poems and Plays of Robert Browning
Mr. Sludge, 'The Medium'

La Experiencia madre es de la ciencia.
[Experiment is the mother of science.]

Cahier, Charles
Quelques Six Mille Proverbes (p. 248)

"This is the *most* interesting Experiment" the Professor announced. "It will need *time*, I'm afraid: but that is a trifling disadvantage. Now observe. If I were to unhook this weight, and let go, it would fall to the ground. You do not deny *that*?"

Nobody denied it.

"And in the same way, if I were to bend this piece of whalebone round the post—thus—and put the ring over this hook—thus—it stays bent: but, if I unhook it, it straightens itself again. You do not deny *that*?"

Again, nobody denied it.

"Well, now suppose we left things as they are, for a long time. The force of the *whalebone* would get exhausted, you know, and it would stay bent, even when you unhooked it. Now, *why* shouldn't the same thing happen with the *weight*. The *whalebone* gets so used to being bent, that it ca'n't *straighten* itself any more. Why shouldn't the *weight* get so used to being held up, that it ca'n't *fall* any more? That's what I want to know!"

"That's what *we* want to know!" echoed the crowd.

"How long must we wait?" grumbled the Emperor.

The Professor looked at his watch. "Well, I *think* a thousand years will do to *begin* with, . . ."

Carroll, Lewis
The Complete Works of Lewis Carroll
Sylvie and Bruno Concluded
Chapter XXI

The statistician who supposes that his main contribution to the planning of an experiment will involve statistical theory, finds repeatedly that he makes his most valuable contribution simply by persuading the investigator to explain why he wishes to do the experiment, by persuading him to justify the experimental treatments, and to explain why it is that the experiment, when completed, will assist him in his research.

Cox, Gertrude M.
Lecture in Washington, 11 January 1951

If you knew some of the experiments (if they may be so-called) which I am trying, you would have a good right to sneer, for they are so absurd even in *my* opinion that I dare not tell you.

Darwin, Charles
The Life and Letters of Charles Darwin
Volume I
C. Darwin to J.D. Hooker
[April 14th, 1855] (p. 415)

WE MUST KNOW MORE ABOUT A PLAN THAN THE PROBABILITIES OF SELECTION. WE MUST KNOW ALSO THE PROCEDURE BY

WHICH TO DRAW THE SAMPLING UNITS, AND THE FORMULA OR PROCEDURE BY WHICH TO CALCULATE THE ESTIMATE.

Deming, William Edwards
Sampling Design in Business Research (p. 39)

Those who fear muddy feet will never discover new paths.

Eldridge, Paul
Maxims for a Modern Man
1286

Do not be too timid and squeamish about your actions. All life is an experiment. The more experiments the better.

Emerson, Ralph Waldo
Journals of Ralph Waldo Emerson

. . . the null hypothesis is never proved or established, but is possibly disapproved, in the course of experimentation. Every experiment may be said to exist only in order to give the facts a chance of disproving the null hypothesis.

Fisher, Sir Ronald A.
The Design of Experiments (p. 19)

To consult the statistician after an experiment is finished is often merely to ask him to conduct a *post mortem* examination. He can perhaps say what the experiment died of.

Fisher, Sir Ronald A.
Sankya
Indian Statistical Congress, ca 1938
Volume 4 (p. 17)

There are some things that are sure to go wrong as soon as they stop going right.

Green, Celia
The Decline and Fall of Science
Aphorisms (p. 171)

No experiment can be more precarious than that of a half-confidence.

Godwin, William
St. Leon; A Tale of the Sixteenth Century (p. 140)

. . . it being justly esteemed an unpardonable temerity to judge the whole course of nature from one single experiment, however accurate or certain.

Hume, David
An Enquiry Concerning Human Understanding
Section VII (p. 77)

Why think? Why not try the experiment?

> **Hunter, John**
> Letter to Edward Jenner, August 2, 1775

Ancient traditions, when tested by the severe processes of modern investigation, commonly enough fade away into mere dreams: but it is singular how often the dream turns out to have been a half-waking one, presaging a reality.

> **Huxley, Thomas H.**
> *Man's Place in Nature*
> I (p. 1)

Hiawatha Designs an Experiment

> **Kendall, Maurice G.**
> *The American Statistician*
> Hiawatha Designs an Experiment
> Volume 13, Number 5, December 1959 (pp. 23–4)

. . . in the full tide of successful experiment . . .

> **Jefferson, Thomas**
> *The Inaugural Addresses of the Presidents of the United States*
> First Inaugural Address at Washington DC, March 4, 1801

. . . theory is a good thing but a good experiment lasts forever.

> **Kapitza, Pyetr Leonidovich**
> *Nature*
> Science East and West: Reflections of Peter Kapitza
> (Book Review by Nevill Mott)
> Volume 288, 11 December 1980 (p. 627)

Every experiment is like a weapon which must be used in its particular way—a spear to thrust, a club to strike. Experimenting requires a man who knows when to thrust and when to strike, each according to need and fashion.

> **Paracelsus, Philippus Aureolus**
> *Surgeon's Book*

If one wishes to obtain a definite answer from Nature one must attack the question from a more general and less selfish point of view.

> **Planck, Max**
> *A Survey of Physics*
> The Unity of the Physical Universe (p. 15)

Polus. O chaerephon, there are many arts among mankind which are experimental, and have their origin in experience, for experience makes the days of men to proceed according to art, and inexperience according to chance, and different persons in different ways are proficient in different arts, and the best persons in the best arts.

Plato
Gorgias
448

Experiment is the sole source of truth. It alone can teach us something new; it alone can give us certainty.

Poincaré, Henri
The Foundations of Science
Science and Hypothesis (p. 127)

It is often said that experiments must be made without preconceived ideas. That is impossible. Not only would it make all experiment barren, but that would be attempted which could not be done.

Poincaré, Henri
The Foundations of Science
Science and Hypothesis (p. 129)

If your experiment needs statistics, you ought to have done a better experiment.

Rutherford, Ernest
Quoted in N.T. Bailey's
The Mathematical Approach to Biology and Medicine
Chapter 2 (p. 23)

Prove all things; hold fast that which is good.

The Bible
1 Thessalonians 5:21

Tuesday. She has taken up with a snake now. The other animals are glad, for she was always experimenting with them and bothering them; and I am glad, because the snake talks, and this enables me to get a rest.

Twain, Mark
Adam's Diary

The Eleven Phases of an Experiment

1. Wild enthusiasm
2. Exciting commitments
3. Total confusion
4. Re-evaluation of goals
5. Disillusionment
6. Cross-accusations
7. Search for the guilty
8. Punish the innocent
9. Promote the non-participants
10. Verbally assassinate visible leaders
11. Write and publish the report

Unknown

Diversity of treatment has been responsible for much of the criticism leveled against the experiment.

Unknown

No experiment is ever a complete failure. It can always be used as a bad example.

Unknown

You must be using the wrong equipment if an experiment works.

Unknown

If an experiment is not worth doing at all, it is not worth doing well.

Unknown

Allow me to express now, once and for all, my deep respect for the work of the experimenter and for his fight to wring significant facts from an inflexible Nature, who says so distinctly "No" and so indistinctly "Yes" to our theories.

Weyl, Hermann
The Theory of Groups and Quantum Mechanics
Introduction (p. xx)

. . . experiment is nothing else than a mode of cooking the facts for the sake of exemplifying the law.

Whitehead, Alfred North
Adventures of Ideas
Foresight
Section I

FACTS

From dreams I proceed to facts.

Abbott, Edwin A.
Flatland (p. 68)

The facts seemed certain, or at least as certain as other facts; all they needed was explanation.

Adams, Henry
The Education of Henry Adams
The Abyss of Ignorance (p. 435)

Entrenching himself behind an undeniable fact.

Alcott, Louisa May
Little Women
XXXV

. . . with a true view all the data harmonize, but with a false one the facts soon clash.

Aristotle
The Nicomachean Ethics
Book I, Chapter VIII

Deny the facts altogether, I think, he hardly can.

Arnold, Matthew
Discourse in America
Literature and Science (p. 101)

"Well facts are facts," said Tilly sulkily.

"So they are, and figures are figures. Stop subtracting the date and get with it."

Balchin, Nigel
The Small Back Room (p. 24)

"Am I supposed to give all the facts, or some of the facts, or my opinions or your opinions or what?"

Balchin, Nigel
The Small Back Room (p. 53)

Facts were never pleasing to him. He acquired them with reluctance and got rid of them with relief. He was never on terms with them until he had stood them on their heads.

Barrie, Sir J.M.
The Greenwood Hat
Love me Never or Forever (pp. 50–51)

To an ordinary person a fact is a fact, and that is all there is to be said about it.

Barry, Frederick
The Scientific Habit of Thought (p. 91)

A fact is no simple thing.

Barry, Frederick
The Scientific Habit of Thought (p. 91)

Facts are to begin with, *coercive.*

Barry, Frederick
The Scientific Habit of Thought (p. 92)

In science one must search for ideas. If there are no ideas, there is no science. A knowledge of facts is only valuable in so far as facts conceal ideas: facts without ideas are just the sweepings of the brain and the memory.

Belinski, Vissarion Grigorievich
Collected Works
Volume 2 (p. 348)

If the facts used as the basis for reasoning are ill-established or erroneous, everything will crumble or be falsified; and it is thus that errors in scientific theories most often originate in errors of fact.

Bernard, Claude
An Introduction to the Study of Experimental Medicine (p. 13)

Facts are neither great or small in themselves.

Bernard, Claude
An Introduction to the Study of Experimental Medicine (p. 34)

A fact is nothing in itself, it has value only through the idea connected with it or through the proof it supplies.

Bernard, Claude
An Introduction to the Study of Experimental Medicine (p. 53)

It is a statistikal fakt, that the wicked work harder tew reach Hell, than the righteous do tew git to heaven.

Billings, Josh
Old Probability: Perhaps Rain—Perhaps Not
April 1870

This plain, plump fact.

Browning, Robert
The Poems and Plays of Robert Browning
Mr. Sludge, 'The Medium'

But facts are facts and flinch not.

Browning, Robert
The Ring and the Book
Part II
Half-Rome, l. 1049

. . . in the long run there is no contending against facts; it is useless to "kick against the pricks".

Buchner, Ludwig
Force and Matter
Preface to the First Edition (p. vi)

But enough of facts!

Buchner, Ludwig
Force and Matter
Brain and Mind (p. 231)

Plain matters of fact are terrible stubborn things.

Budgell, Eustace
Liberty and Progress
ii, 76

Facts are chiels that winna ding an' downa be disputed.
[Facts are entities which cannot be manipulated or disputed.]

Burns, Robert
The Complete Poetical Works of Robert Burns
A Dream, l. 30

I grow to honor facts more and more, and theory less and less. A fact, it seems to me, is a great thing—a sentence printed, if not by God, then at least by the Devil.

Carlyle, Thomas
Letter to Ralph Waldo Emerson, April 29, 1836

First accumulate a mass of Facts: and *then* construct a Theory.

Carroll, Lewis
The Complete Works of Lewis Carroll
Sylvie and Bruno
Queer Street, Number Forty

The Theory hardly rose to the dignity of a Working Hypothesis. Clearly more Facts were needed.

Carroll, Lewis
The Complete Works of Lewis Carroll
Sylvie and Bruno
Queer Street, Number Forty

Some facts are so incredible that they are believed at once, for no one could possibly have imagined them.

Clarke, Arthur C.
The Lost Worlds of 2001
Chapter 30

Every lawyer knows that the name of the game is what label you succeed in imposing on the facts.

Cohen, Jerome
Time
Tense Triangle—What to Do About Taiwan
June 7, 1971 (p. 24)

They demand facts from him, as if facts could explain anything.

Conrad, Joseph
Lord Jim
IV

The language of facts, that are so often more enigmatic than the craftiest arrangement of words.

Conrad, Joseph
Lord Jim
XXXVI

Facts make life long—not years.

Crawford, F. Marion
Don Orsino
XV

The trouble with facts is that there are so many of them.

Crothers, Samuel McChord
The Gentle Reader (p. 183)

Now, what I want are facts . . . Facts alone are wanted in life.

Dickens, Charles
The Work of Charles Dickens
Hard Times
Book I, Chapter I

In this life we want nothing but Facts, sir, nothing but Facts.

Dickens, Charles
The Work of Charles Dickens
Hard Times
Book I, Chapter 1

The labors of others have raised for us an immense reservoir of important facts.

Dickens, Charles
The Work of Charles Dickens
Pickwick Papers
Chapter 4 (p. 46)

Facts and Figures! Put 'em down.

Dickens, Charles
The Work of Charles Dickens
The Chimes: First Quarter

With fuller knowledge we should sweep away the references to probability and substitute the exact facts.

Eddington, Sir Arthur Stanley
The Nature of the Physical World (p. 305)

I am absolutely convinced that one will eventually arrive at a theory in which the objects connected by laws are not probabilities, but conceived facts.

Einstein, Albert
Letter to Max Born
December 3, 1947

We hew and saw and plane facts to make them dovetail with our prejudices, so that they become mere ornaments with which to parade our objectivity.

Eldridge, Paul
Maxims for a Modern Man
2098

Combining superstition with facts is often as efficacious as breaking rocks with fists.

Eldridge, Paul
Maxims for a Modern Man
2159

Facts only emphasize that men are guided by fancies.

Eldridge, Paul
Maxims for a Modern Man
2168

You seem to have a decided faculty for *digesting facts* as evidence.

Eliot, George
The George Eliot Letters
Volume II (p. 205)

Facts are stubborn things.

Eliott, Ebenezer
Field Husbandry (p. 35)

No facts are to me sacred; none are profane; I simply experiment, an endless seeker, with no Past at my back.

Emerson, Ralph Waldo
Essays
Circles (p. 297)

No anchor, no cable, no fences avail to keep a fact a fact.

Emerson, Ralph Waldo
Essays
History (p. 14)

I distrust the facts and the inferences.

Emerson, Ralph Waldo
Essays
Experience (p. 57)

A little fact is worth a whole limbo of dreams . . .

Emerson, Ralph Waldo
Lectures and Biographical Sketches
The Superlative

Facts are not science—as the dictionary is not literature.

Fabing, Harold
Mar, Ray
Fischerisms (p. 21)

We may make our own opinions, but facts were made for us; and, if we evade or deny them, it will be the worse for us.

Froude, James Anthony
Short Studies on Great Subjects
Times of Erasmus, Desderius and Luther (p. 41)

The necessitarian fall back upon the experienced reality of facts.

Froude, James Anthony
Short Studies on Great Subjects
Calvinism (p. 11)

These are facts which no casuistry can explain away.

Froude, James Anthony
Short Studies on Great Subjects
Calvinism (p. 11)

It is through a conviction of the inadequacy of all formulas to cover the facts of nature, it is by a constant recollection of the fallibility of the best instructed intelligence, and by an unintermittent skepticism which goes out of its way to look for difficulties, that scientific progress has been made possible.

Froude, James Anthony
Short Studies on Great Subjects
The Grammar of Assent (pp. 89–90)

Facts are no longer looked in the face, and objections are either ignored altogether or are caricatured in order to be answered.

Froude, James Anthony
Short Studies on Great Subjects
The Grammar of Assent (p. 99)

Facts can be accurately known to us only by the most rigid observation and sustained and scrutinizing skepticism . . .

Froude, James Anthony
Short Studies on Great Subjects
Scientific Method Applied to History (p. 453)

4th VOICE. Let's get the facts. Let's go and watch TV.

Garson, Barbara
MacBird
Act I, Scene VII (p. 18)

Her taste exact
For faultless fact
Amounts to a disease.

Gilbert, W.S.
Sullivan, Arthur
The Complete Plays of Gilbert and Sullivan
The Mikado
Act II

The acts and facts of to-day continually diverge from the concepts of yesterday.

Gilman, Charlotte P.
Human Work
Concept and Conduct (p. 41)

"And of what possible use is that information?" Kerk asked.
"Well, you never know; might come in handy."

Harrison, Harry
Astounding
The Mothballed Spaceship (p. 212)

What are the facts? Again and again and again—what are the facts? Shun wishful thinking, ignore divine revelation, forget what "the stars foretell", avoid opinion, care not what the neighbors think, never mind the unguessable "verdict of history"—what are the facts, and how many decimal places? You pilot always into an unknown future; facts are your single clue. Get the facts!

Heinlein, Robert A.
Time Enough for Love (p. 264)

The more facts one has, the better the judgment one can make, but one must never forget the corollary that the more facts one has, the easier it is to put them together wrong.

Heyworth, Sir Geoffrey
Inaugural Address
President of the Royal Statistical Society
1949

All generous minds have a horror of what are commonly called "facts". They are the brute beasts of the intellectual domain.

Holmes, O.W.
The Autocrat of the Breakfast Table
Chapter 1

Absolute, preemptory facts are bullies and those who keep company with them are apt to get a bullying habit of mind . . .

Holmes, O.W.
The Autocrat of the Breakfast Table
Chapter 3

Facts always yield the place of honor in conversation, to thoughts about facts; but if a false note is uttered, down comes the finger on the key and the man of facts asserts his true dignity.

Holmes, O.W.
The Autocrat of the Breakfast Table
Chapter 6

"The ideal reasoner," he remarked, "would, when he has once been shown a single fact in all its bearings, deduce from it not only all the chain of events which led up to it, but also all the results which would follow from it . . ."

Holmes, Sherlock
in Arthur Conan Doyle's
The Complete Sherlock Holmes
The Five Orange Pips

"I find it hard enough to tackle facts, Holmes, without flying away after theories and fancies."

"You are right," said Holmes demurely; "you do find it very hard to tackle the facts."

Holmes, Sherlock
in Arthur Conan Doyle's
The Complete Sherlock Holmes
The Boscombe Valley Mystery

There is nothing more deceptive than an obvious fact.

Holmes, Sherlock
in Arthur Conan Doyle's
The Complete Sherlock Holmes
The Boscombe Valley Mystery

If you will find the facts, perhaps others may find the explanation.

Holmes, Sherlock
in Arthur Conan Doyle's
The Complete Sherlock Holmes
The Problem of Thor Bridge

A further knowledge of facts is necessary before I would venture to give a final and definite opinion.

Holmes, Sherlock
in Arthur Conan Doyle's
The Complete Sherlock Holmes
The Adventure of Wisteria Lodge, I

"I should have more faith," he said; "I ought to know by this time that when a fact appears to be opposed to a long train of deductions, it invariably proves to be capable of bearing some other interpretation."

Holmes, Sherlock
in Arthur Conan Doyle's
The Complete Sherlock Holmes
A Study in Scarlet

Facts are ventriloquists' dummies. Sitting on a wise man's knee they may be made to utter words of wisdom; elsewhere they say nothing or talk nonsense . . .

Huxley, Aldous
Time Must Have a Stop (p. 301)

Facts do not cease to exist because they are ignored.

Huxley, Aldous
Proper Studies
A Note on Dogma (p. 205)

. . . he had one eye upon fact, and the other on Genesis.

Huxley, Thomas H.
Methods and Results
The Progress of Science (p. 127)

Those who refuse to go beyond fact rarely get as far as fact.

Huxley, Thomas H.
Methods and Results
The Progress of Science (p. 62)

The fatal futility of Fact.

James, Henry
The Spoils of Poynton
Preface

I have to forge every sentence in the teeth of irreducible and stubborn facts.

James, William
Letter to brother Henry James

Just statin' eevidential facts beyon' all argument.

Kipling, Rudyard
Rudyard Kipling's Verse
McAndrew's Verse

An impartial and reliable research substitutes facts for hunches.

Kratovil, Robert
Real Estate Law (p. 419)

Facts are stubborn things.

LaSage, Alan René
The Adventure of Gil Blas of Santillane
Book X, Chapter I

The ultimate umpire of all things in Life is—Fact.

Laut, Agnes C.
The Conquest of the Great Northwest
Part III, Chapter XX (p. 391)

The method of how psychologists as scientists dispose of facts is of special interest. One of the most common is to give the facts a new name. In this way they are given a special compartment and therefore cease to infringe on the privacy of the theory.

Maier, N.R.F.
The American Psychologist
Maier's Law
March 1960 (p. 208)

If the facts do not conform to the theory, they must be disposed of.

Maier, N.R.F.
The American Psychologist
Maier's Law
March 1960 (p. 208)

To all facts there are laws,
The effect has its cause, and I mount to the cause.

Meredith, Owen (Lord Lytton)
Lucile
Part II, canto iii, stanza 8

What you want are facts, not opinions—

Nightingale, Florence
Notes on Nursing
Chapter XIII

Facts are carpet-tacks under the pneumatic tires of theory.

O'Malley, Austin
Keystones of Thought

. . . when technical people talk they always emphasize the facts that they are not sure.

Oppenheimer, Julius Robert
Harper's Magazine
The Tree of Knowledge
Volume 217, October 1958 (p. 57)

When five days later the Morning Star has lifted up its radiance bright from out the ocean waves, then is the time that spring begins. But yet be not deceived, cold days are still in store for thee, indeed they are: departing winter leaves behind great tokens of himself.
[Believe the facts]

Ovid
Fasti
II, l. 149

I'm not afraid of facts. I welcome facts—*but a congeries of acts is not equivalent to an idea.* This is the essential fallacy of the so-called "scientific" mind. People who mistake facts for ideas are incomplete thinkers; they are gossips.

Ozick, Cynthia
Quoted in Francis Klagsbrun's
The First Ms. Reader
We are the Crazy Lady and Other Feisty Feminist Fables (p. 67)

Learn, compare, collect facts.

Pavlov, Ivan
Bequest of the Academic Youth of Soviet Russia
1936

Gross's Postulate. Facts are not all equal. There are good facts and bad facts. Science consists of using good facts.

Peers, John
1001 Logical Laws (p. 35)

Res ipse laquitur
[The fact speaks for itself]

Phrase, Latin

The facts are to blame my friend. We are all imprisoned by facts.

Pirandello, Luigi
The Rules of the Game, The Life I Gave you [and] Lazarus

Nothing is more interesting to the true theorist than a fact which directly contradicts a theory generally accepted up to that time, for this is his particular work.

Planck, Max
A Survey of Physics
New Path of Physical Knowledge (pp. 72–3)

Res ipsa testit.
[Facts speak for themselves.]

Plautus
Aulularia
1, 421
see also Aldous Huxley's
Time Must Have a Stop (p. 301)

The facts of greatest outcome are those we think simple; may be they really are so, because they are influenced only by a small number of well-defined circumstances, may be they take on an appearance of simplicity because the various circumstances upon which they depend obey the laws of chance and so come to mutually compensate.

Poincaré, Henri
The Foundations of Science
Science and Method (pp. 544–5)

. . . the most interesting facts are those which may serve many times; these are the facts which have a chance of coming up again. We have been so fortunate as to be born in a world where there are such.

Poincaré, Henri
The Foundations of Science
Science and Method (p. 363)

Science is built up with facts, as a house is with stones. But a collection of facts is no more a science than a heap of stones is a house.

Poincaré, Henri
The Foundations of Science
Science and Hypothesis (p. 127)

A fact is a fact.

Poincaré, Henri
The Foundations of Science
Science and Hypothesis (p. 128)

I beg to advise you of the following facts of which I happen to be the equally impartial and horrified witness.

Queneau, Raymond
Exercises in Style
Official Letter

But it was a fact—not a theory, not a hypothesis, but a fact—that she was attracted, that she did trust, that she did believe.

Roberts, Nora
Without a Trace
Chapter 5 (p. 104)

With solid facts on hand one may have only one undisputed explanation; with no facts, there can be a dozen argumentative ones.

Romanoff, Alexis L.
Encyclopedia of Thoughts
Aphorisms
2411

Facts were facts, fantasies were fantasies. And never the twain should meet.

Ross, JoAnn
Tempting Fate
Chapter One

Science, as its name implies, is primarily knowledge; by convention it is knowledge of a certain kind, the kind namely, which seeks general laws connecting a number of particular facts.

Russell, Bertrand A.
The Scientific Outlook
Introduction

One of the chief motivations behind the attempt to defend a distinction between theoretical and observational terms has been the desire to explain how a theory can be tested against the data of experience, and how one theory can be said to "account for the facts" better than another; that is, to give a precise characterization of the idea, almost universally accepted in modern times, that the sciences are "based on experience," that they are "empirical".

Shapere, Dudly
Philosophical Problems of Natural Science (p. 15)

Patiokim: In Russia we face facts.

Edstaston: In England, sir, a gentleman never faces any facts if they are unpleasant facts.

Patiokim: In real life, all facts are unpleasant.

Shaw, George Bernard
Complete Plays with Prefaces
Volume IV
Great Catherine
Scene I

A mere fact will never stop an Englishman.

Shaw, George Bernard
Speech, October 28, 1930

. . . the facts, the stubborn, immovable facts.

Smedley, F.E.
Frank Fairlegh or Scenes from the Life of a Private Pupil
Chapter 49

Facts are facts, as the saying is.

Smollett, Tobias
The Life and Adventures of Sir Launcelot Greaves
Chapter III

Comment is free but facts are on expense.

Stoppard, Tom
Night and Day
Act 2

Facts speak louder than statistics.

Streatfield, Mr. Justice Geoffrey
The Observer
Sayings of the Week, 19 March, 1950

Let us look at the facts.

Terence
Adelphoe
l. 796

Matters of fact, which as Mr. Budgell somewhere observes, are very stubborn things.

Tindall, Matthew
The Will of Matthew Tindall (p. 23)

Get your facts first, and then you can distort them as much as you please.

Twain, Mark
Quoted in Rudyard Kipling's
From Sea to Sea
An Interview with Mark Twain

My mind is made up, do not confuse me with facts.

Unknown

We want the facts to fit the preconceptions. When they don't, it is easier to ignore the facts than change the preconceptions.

West, Jessamyn
The Quaker Reader
Introduction (p. 2)

No matter of fact can be mathematically demonstrated, though it may be proved in such a manner as to leave no doubt on the mind.

Whatley, Richard
Logic
IV

It was an ultimate fact.

Whitehead, Alfred North
Science and the Modern World
Chapter III

They remain 'stubborn fact' . . .

Whitehead, Alfred North
Adventures of Ideas
Philosophic Method
Section XVII

But a fact 'contrary' is consciousness in germ . . . Consciousness requires more than the mere entertainment of theory. It is the feeling of the contrast of theory, as *mere* theory with fact, as *mere* fact. This contrast holds whether or not the theory is correct.

Whitehead, Alfred North
Process and Reality
Part II
Discussions and Applications
Propositions
Section I

A chain of facts is like a barrier reef. On one side there is wreckage, and beyond it harbourage and safety.

Whitehead, Alfred North
Process and Reality
Part III
The Theory of Prehensions
The Theory of Feelings
Section IV

There is nothing in the real world which is merely an inert fact . . .

Whitehead, Alfred North
Process and Reality
The Theory of Extension
Part IV

Facts fled before philosophy like frightened forest things.

Wilde, Oscar
The Picture of Dorian Gray
III

Science is built up with facts, as a house is with stones. But a collection of facts is no more a science than a heap of stones is a house.

Henri Poincaré – (See p. 101)

FORECAST

Foreknowledge of the future makes it possible to manipulate both enemies and supporters.

Aron, Raymond
The Opium of the Intellectuals
Chapter IX (p. 284)

How could one haruspex look another in the face without laughing?

Cicero
Cicero: De Senectute, De Amicitia, De Divinatione
De Divinatione
ii, 24

Forecasting in economics is an activity fully licensed in the City of Action and the City of Intellect. Sought and subsidized by executives in government and business, it is also recognized and accredited by the universities. For it to attain so remarkable a status, two suspicions had to be overcome: that of men of action "the speculative views of intellectuals who lack any experience of reality"; and that, even stronger, of men of learning about "intellectual adventurism which discredits science by going beyond the established facts".

de Jouvenel, Bertrand
The Art of Conjecture (p. 179)

Forecasting is very difficult, especially about the future.

Fiedler, Edgar R.
Across the Board
The Three Rs of Economic Forecasting—Irrational, Irrelevant and Irreverent
June 1977

He who lives by the crystal ball soon learns to eat ground glass.

Fiedler, Edgar R.
Across the Board
The Three Rs of Economic Forecasting—Irrational, Irrelevant and Irreverent
June 1977

The moment you forecast you know you're going to be wrong, you just don't know when and in which direction.

Fiedler, Edgar R.
Across the Board
The Three Rs of Economic Forecasting—Irrational, Irrelevant and Irreverent
June 1977

The herd instinct among forecasters makes sheep look like independent thinkers.

Fiedler, Edgar R.
Across the Board
The Three Rs of Economic Forecasting—Irrational, Irrelevant and Irreverent
June 1977

When you know absolutely nothing about the topic, make your forecast by asking a carefully selected probability sample of 300 others who don't know the answer either.

Fiedler, Edgar R.
Across the Board
The Three Rs of Economic Forecasting—Irrational, Irrelevant and Irreverent
June 1977

If you have to forecast, forecast often.

Fiedler, Edgar R.
Across the Board
The Three Rs of Economic Forecasting—Irrational, Irrelevant and Irreverent
June 1977

I know of no way of judging the future but by the past.

Henry, Patrick
Speech at Second Virginia Convention, March 23, 1775

It appears to me a most excellent thing for the physician to cultivate Prognosis; for by foreseeing and foretelling, in the presence of the sick, the present, the past, and the future, and explaining to omissions which patient have been guilty of, he will be the more readily believed to be acquainted with the circumstances . . .

Hippocrates
The Book of Prognostics
1

Nearly every inference we make with respect to any future event is more or less doubtful. If the circumstances are favorable, a forecast may be made with a greater degree of confidence than if the conditions are not so disposed.

Mellor, J.W.
Higher Mathematics for Students of Chemistry and Physics
Probability and Theory of Errors (p. 498)

We are making forecasts with bad numbers, but bad numbers are all we've got.

Penjer, Michael
The New York Times
September 1, 1989

It is far better to foresee even without certainty than not to foresee at all.

Poincaré, Henri
The Foundations of Science
Science and Hypothesis (p. 129)

Qui bene conjiciet, hunc vatem.
[He who guesses right is the prophet.]

Proverb, Greek

A forecast is a forecast is a forecast. What if an important new trend developed? All the possibilities were considered three months ago, and it's too late to discuss any further changes in this year's projections.

Strong, Lydia
Management Review
Sales Forecasting: Problems and Prospects
September 1956

He's fed in enough data for a dozen forecasts—let the electronic brains do the rest. While the THINK machines grind out prophecies, he can relax and contemplate the cosmos.

Strong, Lydia
Management Review
Sales Forecasting: Problems and Prospects
September 1956

His forecasts could have been presented at the deadline date—but he's held it up six weeks waiting for information which will clear up one "crucial" point—crucial only to him.

Strong, Lydia
Management Review
Sales Forecasting: Problems and Prospects
September 1956

Will he ever be able to correlate all these facts into one forecast that makes sense? What does it matter? He's just obtained a new and exclusive figure on discretionary consumer income in Hudson N.Y.—and he's sublimely happy.

Strong, Lydia
Management Review
Sales Forecasting: Problems and Prospects
September 1956

Two plus two is four? Not to this forecaster. He knows the sales manager (who hired him) wants a different answer.

Strong, Lydia
Management Review
Sales Forecasting: Problems and Prospects
September 1956

The charts rustle as the wind murmurs through the sacred grove. The high priest interprets the prophecy to the waiting supplicant. Business will improve, he says . . . unless it takes a turn for the worse.

Strong, Lydia
Management Review
Sales Forecasting: Problems and Prospects
September 1956

Why fool around with market research? Why try to correlate economic indicators? The correct prediction will strike suddenly—like a bolt from the blue.

Strong, Lydia
Management Review
Sales Forecasting: Problems and Prospects
September 1956

"You've got a tough job ahead of you," the manager told the new employee in the research department. "Our president respected the guy you're replacing and had great faith in his forecasting abilities."

"Was he a statistician?" the employee asked.

"In a way. He used to hang around the lunchroom and read coffee grounds."

Thomsett, Michael C.
The Little Black Book of Business Statistics (p. 140)

It is said that the present is pregnant with the future.

Voltaire
The Portable Voltaire
Philosophical Dictionary
Concatenation of Events

Men have always valued the ability to predict future events, for those who can predict events can guard against them.

Walker, Marshall
The Nature of Scientific Thought (p. 2)

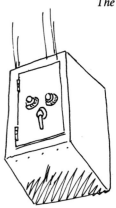

GAMBLING

Gambling is increasing beyond what you could imagine. 'Pitch-and-toss' is too dull: all must bet; women as well as men. Bookies stand about and meet men as they go to and from their work.

Booth, Charles
Charles Booth's London (p. 336)

In moderation, gambling possesses undeniable virtues. Yet it presents a curious spectacle replete with contradictions. While indulgence in its pleasures has always lain beyond the pale of fear of Hell's fires, the great laboratories and respectable insurance palaces stand as monuments to a science originally born of the dice cup.

Kasner, Edward
Newman, James
Mathematics and the Imagination (p. 239)

People don't like to choose #1 in a lottery. 'Choose it,' Reason cries loudly. 'It has as good a chance of winning the 12,000 thalers as any other.' 'In Heaven's name don't choose it,' *a je ne sais quoi* whispers. 'There's no example of such little numbers being listed before great winnings.' And actually no one takes it.

Lichtenberg, Georg
Lichtenberg: Aphorisms & Letters
Aphorisms (p. 46)

There are three roads to ruin; women, gambling and technicians. The most pleasant is with women, the quickest is with gambling, but the surest is with technicians.

Pompidou, Georges
Sunday Telegraph
26 May 1968

He felt the table was having a run of bad luck, but he knew. Gronevelt would never accept that explanation. Gronevelt believed that the house could not lose over the long run, that the laws of percentage were not subject to chance. As gamblers believed mystically in their luck so Gronevelt believed in percentages.

Puzo, Mario
Fools Die: A Novel
Chapter 17 (pp. 187–8)

YOU SMILED !
NO I NEVER- YOU DID !
NO I DIDN'T - YOU DID !
THERE - A SMILE - YOU SMILED !
DID NOT !
DID TOO !
DID NOT ! ETC..!

How could one haruspex look another in the face without laughing?
Cicero – (See p. 106)

GRAPHICS

Every picture tells a story.

Advertisement for Doan's Backache Kidney Pills

One picture is worth ten thousand words.

Advertisement for Royal Baking Powder
Printers Ink
Volume 138, 10 March 1927

When graphing a function, the width of the line should be inversely proportional to the precision of the data.

Albinak, Marvin J.
Quoted in Paul Dickson's
The Official Explanations (p. A-3)

"I'll give you a graphic display," Gerhard said. He punched buttons, wiping the screen. After a moment, cross-hatching for a graph appeared and the points began to blink on . . .

Crichton, Michael
The Terminal Man
Chapter 5 (p. 121)

You can draw a lot of curves through three graph points. You can extrapolate it a lot of ways.

Crichton, Michael
The Terminal Man
Chapter 5 (p. 155)

The preliminary examination of most data is facilitated by the use of diagrams. Diagrams prove nothing, but bring outstanding features readily to the eye; they are therefore no substitutes for such critical tests as may be applied to the data, but are valuable in suggesting such tests, and in explaining the conclusions founded upon them.

Fisher, Sir Ronald A.
Statistical Methods For Research Workers (p. 27)

. . . no nation ranks higher in its collective passion for statistics. In Japan, statistics are the subject of holidays, local and national conventions, award ceremonies and nationwide statistical collection and graph-drawing contests.

Malcolm, Andrew H.
New York Times
Data-Loving Japanese Rejoice on Statistics Day
October 26, 1977. A-1

It pays to keep wide awake in studying any graph. The thing looks so simple, so frank, and so appealing that the careless are easily fooled.

Moroney, M.J.
Facts from Figures
The Magic Lantern Technique (p. 27)

Despite the prevailing use of graphs as metaphors for communicating and reasoning about dependencies, the task of capturing informational dependencies by graphs is not at all trivial.

Pearl, Judea
Probabilistic Reasoning in Intelligent Systems (p. 81)

As to the propriety and justness of representing sums of money, and time, by parts of space, tho' very readily agreed to by most men, yet a few seem to apprehend there may possibly be some deception in it, of which they are not aware . . .

Playfair, William
The Commercial and Political Atlas

A picture is worth more than ten thousand words.

Proverb, Chinese

You must never tell a thing. You must illustrate it. We learn through the eye and not the noggin.

Rogers, Will
The Will Rogers Book
June 25, 1933 (p. 121)

Dost thou love pictures?

Shakespeare, William
The Complete Works of William Shakespeare
The Taming of the Shrew
Introduction, Scene 2, l. 51

Graphical integrity is more likely to result if these six principles are followed:

The representation of numbers, as physically measured on the surface of the graphic itself, should be directly proportional to the numerical quantities represented.

Clear, detailed, and thorough labeling should be used to defeat graphical distortion and ambiguity. Write out explanations of the data on the graphic itself. Label important events in the data.

Show data variations, not design variations.

In time-series displays of money, deflated and standardized units of monetary measurements are nearly always better than nominal units.

The number of information-carrying (variable) dimensions depicted should not exceed the number of dimensions in the data.

Graphics must not quote data out of context.

Tufte, Edward R.
The Visual Display of Quantitative Information (p. 77)

Excellence in statistical graphics consists of complex ideas communicated with clarity, precision, and efficiency. Graphical displays should

- show the data
- induce the viewer to think about the substance rather than about the methodology, graphic design, the technology of graphic production, or something else
- avoid distorting what the data have to say
- present many numbers in a small space
- make large data sets coherent
- encourage the eye to compare different pieces of data
- reveal the data at several levels of detail, from a broad overview to the fine structure
- serve a reasonable clear purpose: description, exploration, tabulation, or decoration
- be closely integrated with the statistical and verbal descriptions of a data set.

Tufte, Edward R.
The Visual Display of Quantitative Information (p. 13)

Of course statistical graphics, just like statistical calculations, are only as good as what goes into them. An ill-specified or preposterous model or a puny data set cannot be rescued by a graphic (or by calculation), no matter how clever or fancy. A silly theory means a silly graphic.

Tufte, Edward R.
The Visual Display of Quantitative Information (p. 15)

A sketch tells me as much in a glance as a dozen pages of print.

Turgenev, Ivan
Fathers and Sons
Chapter 16

WHAT THE ARTIST IS OBVIOUSLY TRYING TO COMMUNICATE TO US THROUGH THIS PICTURE IS THE ANGST HE FELT WHEN TOLD THAT POSSIBLY HE SHOULD TAKE UP BASKETWORK INSTEAD..

Every picture tells a story.
Advertisement for Doan's Backache Kidney Pills – (See p. 113)

HYPOTHESES

Jolie hypothèse elle explique tant de choses.
[A pretty hypothesis which explains many things.]

Asquith, Herbert
Speech in House of Commons
March 29, 1917

. . . hypothetical questions get hypothetical answers.

Baez, Joan
Daybreak
What Would You Do If (p. 134)

Hypothesis, however, is an inference based on knowledge which is insufficient to prove its high probability.

Barry, Frederick
The Scientific Habit of Thought
The Elements of Theory (p. 164)

The shrewd guess, the fertile hypothesis, the courageous leap to a tentative conclusion—these are the most valuable coin of the thinker at work.

Bruner, Jerome Seymour
The Process of Education (p. 14)

"Would you tell me, please, which way I ought to go from here?"

"That depends a good deal on where you want to get to," said the Cat.

"I don't much care where—" said Alice.

"Then it doesn't matter which way you go," said the Cat.

Carroll, Lewis
The Complete Works of Lewis Carroll
Pig and Pepper

117

There is . . . no genuine progress in scientific insight through the Baconian method of accumulating empirical facts without hypotheses or anticipation of nature. Without some guiding idea we do not know what facts to gather . . . we cannot determine what is relevant and what is irrelevant.

Cohen, Morris R.
A Preface to Logic (p. 148)

Since the newness of the hypotheses of this work—which sets the earth in motion and puts an immovable sun at the center of the universe—has already received a great deal of publicity, I have no doubt that certain of the savants have taken grave offense and think it wrong to raise any disturbance among liberal disciplines which have had the right set-up for a long time now.

Copernicus, Nicolaus
On the Revolutions of the Heavenly Spheres
Introduction

But suspicion is a thing very few people can entertain without letting the hypothesis turn, in their minds, into fact.

Cort, David
Social Astonishments
Believing in Books

A false hypothesis, if it serve as a guide for further enquiry, may, at the right stage of science, be as useful as, or more useful than, a truer one for which acceptable evidence is not yet at hand.

Dampier-Whetham, William
Science and the Human Mind
Science in the Ancient World (p. 39)

An honorable man will not be bullied by a hypothesis.

Evans, Bergen
The Natural History of Nonsense
A Tale of a Tub

We see what we want to see, and observation conforms to hypothesis.

Evans, Bergen
The Natural History of Nonsense
A Tale of a Tub

Many confuse hypothesis and theory. An hypothesis is a possible explanation; a theory, the correct one.

Fabing, Harold
Mar, Ray
Fischerisms (p. 7)

In the complete absence of any theory of the instincts which would help us to find our bearings, we may be permitted, or rather, it is incumbent upon us, in the first place to work out any hypothesis to its logical conclusion, until it either fails or becomes confirmed.

Freud, Sigmund
On Narcissism

If the fresh facts which come to our knowledge all fit themselves into the scheme, then our hypothesis may gradually become a solution.

Holmes, Sherlock
in Arthur Conan Doyle's
The Complete Sherlock Holmes
The Adventure of Wisteria Lodge

. . . it is the first duty of a hypothesis to be intelligible . . .

Huxley, Thomas H.
Man's Place in Nature
II (p. 126)

The great tragedy of Science—the slaying of a beautiful hypothesis by an ugly fact.

Huxley, Thomas H.
Collected Essays
Biogenesis and Abiogenesis

This is called the inductive method. Hypothesis, my dear young friend, established itself by a cumulative process; or, to use popular language, if you make the same guess often enough it ceases to be a guess and becomes a Scientific Fact.

Lewis, C.S.
The Pilgrim's Regress: An Allegorical Apology for Christianity, Reason and Romanticism
Book Two
Chapter I (p. 37)

It is a good morning exercise for a research scientist to discard a pet hypothesis every day before breakfast. It keeps him young.

Lorenz, Konrad
On Aggression (p. 12)

We are to admit no more causes of natural things than such as are both true and sufficient to explain their appearances.

Newton, Sir Isaac
Mathematical Principles of Natural Philosophy
Book III, Rule I

In experimental philosophy we are to look upon propositions inferred by general induction from phenomena as accurately or very nearly true, notwithstanding any contrary hypotheses that may be imagined, till such time as other phenomena occur, by which they may either be made more accurate, or liable to exceptions.

Newton, Sir Isaac
Mathematical Principles of Natural Philosophy
Book III, Rule IV

I frame no hypotheses; for whatever is not deduced from the phenomena is to be called an hypothesis; and hypotheses, whether metaphysical or physical, whether of occult qualities or mechanical, have no place in experimental philosophy.

Newton, Sir Isaac
Mathematical Principles of Natural Philosophy
Book III, General Scholium

For sometimes an obvious absurdity follows from its negation, and then the hypothesis is true and certain; or an obvious absurdity follows from its affirmation, and then the hypothesis is considered false; and when we have not yet been able to draw an absurdity either from its negation or from its affirmation, the hypothesis remains doubtful. So that to establish the truth of an hypothesis it is not enough that all the phenomena should follow from it, whereas if there follows from it something opposed to a single phenomenon, that is enough to make certain its falsity.

Pascal, Blaise
Scientific Treatises
Concerning the Vacuum

It is the nature of an hypothesis, when once a man has conceived it, that it assimilates every thing to itself, as proper nourishment; and, from the first moment of your begetting it, it generally grows stronger by every thing you see, hear, read, or understand. This is of great use.

Sterne, Laurence
Tristram Shandy
Book 2, Chapter 19

"I just finished up in the budget review meeting," an exhausted manager told a friend. "It was tough. We were way off on our projections, and I had to explain why."

"How did you do?" the friend asked.

"At first, I tried to tell them we simply made a mistake, but they wouldn't accept that explanation. So then I said that our hypothesis had not included the entire scope of probabilities, and the outcome fell outside of the range we had used."

"And?"

"They loved it."

Thomsett, Michael C.
The Little Black Book of Business Statistics (p. 164)

[Hypothesis] Something murdered by facts.

Unknown

IMPOSSIBLE

It is impossible to import things into an infinite area, there being no outside to import things in from.

Adams, Douglas
The Original Hitchhiker Radio Scripts
Fit the Fifth (p. 101)

What is convincing though impossible should always be prefered to what is possible and unconvincing.

Aristotle
The Poets
Chapter XXIV

Events with a sufficiently small probability never occur, or at least we must act, in all circumstances, as if they were *impossible*.

Borel, Emile
Probabilities and Life
Introduction (pp. 2–3)

Alice laughed. "There's no use trying," she said "one *ca'n't* believe impossible things."

"I daresay you havn't had much practice," said the Queen. "When I was your age, I always did it for half-an-hour a day. Why, sometimes I've believed as many as six impossible things before breakfast . . ."

Carroll, Lewis
The Complete Works of Lewis Carroll
Through the Looking Glass
Wool and Water

A round square or a wooden iron is an absurdity and consequently an impossibility . . .

> **Chestov, Leon**
> *Forum Philosophicum*
> Look Back and Struggle
> Volume 1, Number 1, 1930 (p. 112)

The only way of finding the limits of the possible is by going beyond them into the impossible.

> **Clarke, Arthur C.**
> *The Lost Worlds of 2001*
> Chapter 34

I'll tell you in two words—im-possible.

> **Goldwyn, Samuel**
> *New York Times*
> Obituary, February 1, 1974

Except under controlled conditions, or in circumstances where it is possible to ignore individuals and consider only large numbers and the law of averages, any kind of accurate foresight is impossible.

> **Huxley, Aldous**
> *Time Must Have a Stop* (p. 296)

. . . so many things are possible just as long as you don't know they're impossible.

> **Juster, Norton**
> *The Phantom Tollbooth* (p. 247)

Well, I'll have her: and if it be a match, as nothing is impossible—.

> **Shakespeare, William**
> *The Complete Works of William Shakespeare*
> The Two Gentlemen of Verona
> Act III, Scene 2, l. 379

A likely impossibility is always preferable to an unconvincing possibility.

> **Sheynin, O.B.**
> *Archive for History of Exact Science* (p. 101)

The fact is certain because it is impossible.

> **Tertullian**
> *De Carne Christi*
> Chapter V, Part II

All things, as we know, are impossible in this most impossible of all impossible worlds.

Thurber, James
Lanterns and Lances
The Last Clock

Man can believe the impossible, but man can never believe the improbable.

Wilde, Oscar
Epigrams: Phrases and Philosophies for the Use of the Young
Sebastian Melmoth

INFINITE

The ignorant suppose that infinite number of drawings require an infinite amount of time; in reality it is quite enough that time is infinitely subdivisible, as is the case in the famous parable of the Tortoise and the Hare. This infinitude harmonizes in an admirable manner with the sinuous numbers of Chance and of the Celestial Archetype of the Lottery, adored by the Platonists.

Borges, Jorge Luis
Ficciones
The Babylon Lottery

It is a pity, therefore, that the authors have confined their attention to the relatively simple problem of determining the approximate distribution of arbitrary criteria and have failed to produce any sort of justification for the tests they propose. In addition to those functions studied there are an infinity of others, and unless some principle of selection is introduced we have nothing to look forward to but an infinity of test criteria and an infinity of papers in which they are described.

Box, G.E.P.
Journal of the Royal Statistical Society
Discussion
Series B, 18, 1956 (p. 29)

I had expressed my wish to have a *thermometer of probability*, with impossibility at one end, as 2 plus 2 makes 5, and necessity at the other as 2 plus 2 make 4.

de Morgan, Augustus
Budget of Paradoxes
Volume II
James Smith Once More (p. 247)

KNOWLEDGE

Incomplete knowledge must be considered as perfectly normal in probability theory; we might even say that, if we knew all the circumstances of the phenomena, there would be no place for probability, and we would know the outcome with certainty.

Borel, Emile
Probability and Certainty
Chapter I (p. 13)

Thus, the scientist must recognize the statistical aspect of much of his knowledge, not, on the one hand, unduly hesitating to accept it as true if the probability is reasonably high, but, on the other hand, maintaining an alertness to the possibility that what may for good appear to be highly improbable may indeed occur or be true.

Fischer, Robert B.
Science, Man and Society (p. 37)

I am convinced that it is impossible to expound the methods of induction in a sound manner, without resting them on the theory of Probability. Perfect knowledge alone can give certainty, and in nature perfect knowledge would be infinite knowledge, which is clearly beyond our capacities. We have, therefore, to content ourselves with partial knowledge,—knowledge mingled with ignorance, producing doubt.

Jevons, W.S.
The Principles of Science
Chapter 10 (p. 197)

We give them an excellent survey of the methods and techniques of thinking, taken from logic, statistics, scientific method, psychology, and mathematics.

Skinner, B.F.
Walden Two (p. 111)

LAWS

But physicians have nothing to do with what is called the *law of large numbers*, a law which, according to a great mathematician's expression, is always true in general and false in particular.

Bernard, Claude
An Introduction to the Study of Experimental Medicine (p. 138)

Negative expectations yield negative results.
Positive expectations yield negative results.

Bloch, Arthur
Murphy's Law
The Nonreciprocal Laws of Expectation (p. 21)

Indeed, the laws of chance are just as necessary as the causal laws themselves.

Bohm, D.
Causality and Chance in Modern Physics (p. 23)

[in quantum mechanics] we have the paradoxical situation that observable events obey laws of chance, but that the probability for these events itself spreads according to laws which are still in all essential features causal laws.

Born, Max
Natural Philosophy of Cause and Chance (p. 103)

. . . if they do only one jump, you know, there's a fifty percent chance of an injury. Two jumps it's eighty percent. The third time, it's dead certain they won't get off scot free. You see? It's not a question of training, but the law of averages.

Boulle, Pierre
The Bridge over the River Kwai
Part Two, Chapter 8 (p. 67)

In the course of the committee's investigations, it had been discovered, to everyone's dismay, that the Law of Averages had never been incorporated into the body of federal jurisprudence, and though the upholders of States' Rights rebelled violently, the oversight was at once corrected, both by Constitutional amendment and by a law—the Hills–Slooper Act—implementing it. According to the Act, people were *required* to be average, and, as the simplest way of assuring it, they were divided alphabetically and the permissible activities catalogued accordingly.

Coates, Robert M.
The World of Mathematics
Volume 3
The Law (p. 2271)

I believe neither in chance nor in miracle, but only in phenomena regulated by laws.

de Jouvenel, Bertrand
Quoted in Ludwig Buchner's
Force and Matter (p. 80)

It would be splendid if all action required in social, economic, and industrial planning could be based on scientific laws; but actually, so many of the laws remain yet to be discovered that most action must be taken on the basis of knowledge of the subject matter in related fields.

Deming, William Edwards
Statistical Adjustment of Data (p. 11)

Ashley–Perry Statistical Axioms

(1) Numbers are tools, not rules.

(2) Numbers are symbols for things; the numbers and the things are not the same.

(3) Skill in manipulating numbers is a talent, not evidence of divine guidance.

(4) Like other occult techniques of divination, the statistical method has a private jargon deliberately contrived to obscure its methods from nonpractitioners.

(5) The product of an arithmetical computation is the answer to an equation; it is not the solution to a problem.

(6) Arithmetical proofs of theorems that do not have arithmetical bases prove nothing.

Dickson, Paul
The Official Rules (p. A-5)

The Greeks, says Mr. Galton, if they had known of the law of errors, would have personified and deified it; the moderns should at least respect it as the most universal law of nature.

> **Edgeworth, Francis Ysidro**
> *Journal of the Royal Statistical Society*
> On the Representation of Statistics by Mathematical Formula (concluded)
> Volume XLII, 1899 (p. 552)

As far as the laws of mathematics refer to reality, they are not certain; and as far as they are certain, they do not refer to reality.

> **Einstein, Albert**
> *Sidelights on Relativity*
> Geometry and Experience (p. 28)

But when I came to reflect on the facts observed, I was struck by their singularity. Moustache hairs are shed very freely, but they do not drop out at regular intervals. One, two, or more hairs in any one box would not have been surprising. A man who was in the habit of pulling or stroking his moustache might dislodge two or three at once. The surprising thing was the regularity with which these hairs occurred; one, and usually one only, in each box, and no complete box in which there was none. It was totally opposed to the laws of probability.

> **Freeman, R. Austin**
> *A Certain Dr. Thorndyke*
> Thorndyke Connects the Links

Superstition, heroworship, ignorance of the laws of probability, religious, political, or speculative prejudice. One or other of these has tended from the beginning to give us distorted pictures.

> **Froude, James Anthony**
> *Short Studies on Great Subjects*
> Scientific Method Applied to History (p. 470)

It will, I trust, be clearly understood that the numbers of men in the several classes in my table depend on no uncertain hypothesis. They are determined by the assured law of deviations from the average.

> **Galton, Francis**
> *Hereditary Genius*
> According to Their Natural Gift (p. 30)

I know of scarcely anything so apt to impress the imagination as the wonderful form of cosmic order expressed by the "Law of Frequency of Error". The law would have been personified by the Greeks and deified, if they had known of it. It reigns with serenity, and in complete effacement amidst the wildest confusion. The huger the mob, and the greater the apparent anarchy, the more perfect is its sway. It is the supreme law of Unreason. Whenever a large sample of chaotic elements are taken in hand and marshalled in the order of their magnitude, an unsuspected and most beautiful form of regularity proves to have been latent all along. The tops of the marshalled now form a flowing curve of invariable proportions; and each element, as it is sorted in place, finds, as it were, a preordained niche, accurately adapted to fit it.

Galton, Francis
Natural Inheritance
Normal Variability (p. 66)

. . . but the laws of probability, so true in general, so fallacious in particular.

Gibbon, Edward
Gibbon's Autobiography (p. 124)

"Law" means a rule which we have always found to hold good, and which we expect always will hold good.

Huxley, Thomas H.
Collected Essays
On Descartes' "Discourse Touching the Method of Using One's Reason
Rightly and of Seeking Scientific Truth"
Volume I

. . . all the richness of structure observed in the natural world is not a consequence of the complexity of physical law, but instead arises from the many-times repeated application of quite simple laws.

Kadanoff, Leo P.
Physics Today
Complete Structure from Simple Systems
March 1991 (p. 9)

. . . laws serve to explain events and theories to explain laws; a good law allows us to predict new facts and a good theory new laws. At any rate, the success of prediction . . . adds credibility to the beliefs which led to it, and a corresponding force to the explanations they provide.

Kaplan, Abraham
The Conduct of Inquiry
Chapter IX, Section 40 (p. 346)

Dieselbe Ordnung waltet überall:
Im wechselvollen Reigen der Gestirne
Gebietet das Gesetz nach Mass und Zahl,
Wie in des Menschen denkendem Gehirne.

[The same order rules everywhere;
the law of measure and number rules
in the changeful hosts of the stars
as it does in man's thinking brain.]

Krass, F.
Quoted in Ludwig Buchner's
Force and Matter (p. 103)

Law of Probable Disposal: Whatever hits the fan will not be evenly distributed.

Logical Machine Advertisement
Quoted in Paul Dickson's
The Official Rules (p. P-152)

I feel like a fugitive from th' law of averages.

Mauldin, Bill (William Henry)
Up Front
Cartoon caption (p. 39)

We can never achieve absolute truth but we can live hopefully by a system of calculated probabilities. The law of probability gives to natural and human sciences—to human experience as a whole—the unity of life we seek.

Meyer, Agnes
Education for a New Morality (p. 21)

The Law of Causation, the recognition of which is the main pillar of inductive science, is but the familiar truth, that the invariability of succession is found by observation to obtain between every fact in nature and some other fact which has preceded it.

Mill, John Stuart
System of Logic
Book III, Chapter V, Section 2

Osborn's Law. Variables won't, constants aren't.

Osborn, Don
Quoted in Paul Dickson's
The Official Rules (p. O-138)

The purpose I mean is, to show what reason we have for believing that there are in the constitution of things fixed laws according to which events happen . . .

Price, Richard
Introduction to Bayes' *Essays*

When any principle, law, tenet, probability, happening, circumstance, or result can in no way be directly, indirectly, empirically, or circuitously proven, derived, implied, inferred, induced, deduced, estimated, or scientifically guessed, it will always for the purpose of convenience, expediency, political advantage, material gain, or personal comfort, or any combination of the above, or none of the above, be unilaterally and unequivocally assumed, proclaimed, and adhered to as absolute truth to be undeniably, universally, immutably, and infinitely so, until such time as it becomes advantageous to assume otherwise, maybe.

Rhodes, Charles E.
Quoted in Paul Dickson's
The Official Explanations (p. R-192)

Scientific laws, when we have reason to think them accurate, are different in form from the common-sense rules which have exceptions: they are always, at least in physics, either differential equations, or statistical averages. It might be thought that a statistical average is not very different from a rule with exceptions, but this would be a mistake. Statistics, ideally are accurate laws about large groups; they differ from other laws only in being about groups, not about individuals. Statistical laws are inferred from particular statistics, just as other laws are inferred from particular single occurrences.

Russell, Bertrand A.
The Analysis of Matter
Data, Inferences, Hypotheses, and Theories (p. 191)

I come now to the statistical part of physics, which is concerned with the study of large aggregates. Large aggregates behave almost exactly as they were supposed to do before quantum theory was invented, so that in regard to them the older physics is very nearly right. There is, however, one supremely important law which is only statistical; this is the second law of thermodynamics. It states, roughly speaking, that the world is growing continuously more disorderly.

Russell, Bertrand A.
The Scientific Outlook
Scientific Metaphysics (p. 92)

Only by reducing this element of free will to the infinitesimal, that is, by regarding it as an infinitely small quantity, can we convince ourselves of the absolute inaccessibility of the causes, and then instead of seeking causes, history will take the discovery of laws as its problem.

Tolstoy, Leo
War and Peace
Second Epilogue, Chapter XI

We must discover the laws on which our profession rests, and not invent them.

Unknown

Twyman's Law states that any figure that looks interesting or different is usually wrong.

Unknown

If the law states a precise result, almost certainly it is not precisely accurate; and thus even at the best the result, precisely as calculated, is not likely to occur.

Whitehead, Alfred North
An Introduction to Mathematics
Chapter 3

Laws are statements of observed facts.

Whitehead, Alfred North
Adventures of Ideas
Laws of Nature
Section VII

LIKELIHOOD

"I wonder how we can account for such parallelism in door design," Ted said. "The likelihood of its occurring by chance is astronomically small. Why, this door is the perfect size and shape for human beings!"

Crichton, Michael
Sphere
The Door (p. 64)

Mopworth always took a seat at a window already cracked and taped, or patched with cardboard, banking on the Law of Probability to reduce the likelihood of another rock coming in that one again before it came in another.

de Vries, Peter
Ruben, Ruben
Mopworth
Chapter Thirty-Four (p. 411)

He was a strange boy to be sure. There was always some ground of probability and likelihood mingled with his absurd behaviour. That was the best of it.

Dickens, Charles
The Work of Charles Dickens
Martin Chuzzlewit
Chapter XI (p. 166)

No doubt if it had been discovered who wrote the "Vestiges," many an ingenious structure of probabilities would have been spoiled, and some disgust might have been felt for a real author who made comparatively so shabby an appearance of likelihood.

Eliot, George
Theophrastus Such
The Wasp Credited with the Honeycomb (p. 82)

I have no objection to the study of likelihood as such.

Jefferys, Harold
Proceedings of the Royal Statistical Society
Probability and Scientific Method
Series A, Volume 146, 1934

A professor's enthusiasm for teaching introductory courses varies inversely with the likelihood of his having to do it.

Martin, Thomas L., Jr.
Malice in Blunderland
Fuglemenship (p. 103)

There was not much likelihood now that a third encounter would take place, and the fact is that from that day to this I have never seen the young man again, in conformity with the established laws of probability.

Queneau, Raymond
Exercises in Style
Probabilist

A meeting was called to review the result of a recent market sample for a new product. The president started out by asking, "Will we make a profit?"

The manager of the research department answered, "Based on the specific assumptions applied to our test, there is a reasonable likelihood that response will fall within our range of expectations."

The president leaned over and whispered to his secretary, "What was the answer?"

She whispered back, "Yes."

Thomsett, Michael C.
The Little Black Book of Business Statistics (p. 194)

The likelihood of a thing happening is inversely proportional to its desirability.

Wright, Jim
The Dallas Morning News
September 9, 1969

MEASUREMENT

. . . we must remember that measures were made for man and not man for measures.

Asimov, Isaac
Of Time and Space and Other Things
Part II
Of Other Things (p. 143)

One grain of wheat does not constitute a pile, nor do two grains, nor three and so on. On the other hand, everyone will agree that a hundred million grains of wheat do form a pile. What, then is the threshold number? Can we say that 325,647 grains of wheat do not form a pile, but that 325,648 grains do? If it is impossible to fix a threshold number, it will also be impossible to know what is meant by a pile of wheat; the words can have no meaning, although, in certain extreme cases, everybody will agree about them.

Borel, Emile
Probability and Certainty
Chapter 8 (p. 98)

In every thing, I woot, ther lyth mesure.

Chaucer, Geoffrey
Troylus and Cryseyde
Book ii, l. 715

. . . polynomials are notoriously untrustworthy when extrapolated.

Cochran, William
Cox, Gertrude
Experimental Designs (p. 336)

It is important to realize that it is not the one measurement, alone, but its relation to the rest of the sequence that is of interest.

Deming, William Edwards
Statistical Adjustment of Data (p. 3)

Insistence upon numerical measurement when it is not inherently required by the consequence to be effected, is a mark of respect for the ritual of scientific practice at the expense of its substance.

Dewey, John
Logic: The Theory of Inquiry
Chapter XI (p. 205)

Measurement has meaning only if we can transmit the information without ambiguity to others.

Fox, Russell
Gorbuny, Max
Hooke, Robert
The Science of Science
A Standard Language (p. 31)

Who hath measured the waters in the hollow of his hand . . .?

The Bible
Isaiah 40:12

One of the subjects of Kinsey's study of sexual behavior in the human male afterwards complained bitterly of the injury to his masculine ego. "No matter what I told him," he explained, "he just looked me straight in the eye and asked, 'How many times?' " . . . The principle, "Let's get it down to something we can count!" does not always formulate the best research strategy.

Kaplan, Abraham
The Conduct of Inquiry
Chapter V, Section 20 (p. 171)

Measurement, we have seen, always has an element of error in it. The most exact description or prediction that a scientist can make is still only approximate. If, as sometimes happens, a perfect correspondence with observation does appear, it must be regarded as accidental, and, as Jevons [see *The Principles of Science*, p. 457] . . . remarks, it "should give rise to suspicion rather than to satisfaction".

Kaplan, Abraham
The Conduct of Inquiry
Chapter VI, Section 25 (p. 215)

Proleptically, I would say that whether we can measure something depends, not on that thing, but on how we have conceptualized it, on our knowledge of it, above all on the skill and ingenuity which we can bring to bear on the process of measurement which our inquiry can put to use.

Kaplan, Abraham
The Conduct of Inquiry
Chapter V, Section 20 (p. 176)

We are committed to the scientific method and measurement is the foundation of that method; hence we are prone to assume that whatever is measurable must be significant and that whatever cannot be measured may as well be disregarded.

Krutch, Joseph Wood
Human Nature and Human Condition
Chapter 5 (p. 78)

We are ourselves the measure of the miraculous; if we should find a universal measure, the miraculous elements would disappear, and all things would be of equal size.

Lichtenberg, Georg
Lichtenberg: Aphorisms & Letters
Aphorisms (p. 27)

Coomb's Law. If you can't measure it, I'm not interested.

Peter, Lawrence J.
Human Behavior
Peter's People
August, 1976 (p. 9)

Beauty had been born, not, as we so often conceive it nowadays, as an ideal of humanity, but as *measure*, as the reduction of the chaos of appearances to the precision of linear symbols. Symmetry, balance, harmonic division, mated and mensurated intervals—such were its abstract characteristics.

Read, Herbert
Icon and Idea: The Function of Art in the Development of Human Consciousness
Chapter IV (p. 75)

Crude measurement usually yields misleading, even erroneous conclusions no matter how sophisticated a technique is used.

Reynolds, H.T.
Analysis of Nominal Data (p. 56)

Measurement demands some one–one relations between the numbers and magnitudes in question—a relation which may be direct or indirect, important or trivial, according to circumstances.

Russell, Bertrand A.
The Principles of Mathematics
Entry 164

Measure for Measure

Shakespeare, William
Title of play

Nay, if these measures give any ground of confidence, we think that thy design is not amiss.

Sophocles
The Plays of Sophocles
Trachiniae
l. 587

. . . great as may be the potency of this [the experimental method], or of the preceding methods, there is yet another one so vital that, if lacking it, any study is thought by many authorities not to be scientific in the full sense of the word. This further and crucial method is that of measurement, . . .

Spearman, Charles
Psychology Down the Ages
Volume I (p. 89)

I often say that when you can measure what you are speaking about, and express it in numbers, you know something about it; but when you cannot measure it, when you cannot express it in numbers, your knowledge is of a meager and unsatisfactory kind: it may be the beginning of knowledge, but you have scarcely, in your thoughts, advanced to the stage of *science* whatever the matter might be.

Thompson, William (Lord Kelvin)
Popular Lectures and Addresses (p. 80)

What the measurements will not do, is to get you out of the crisis you are already in.

Unknown

If you don't measure it, it won't happen.

Unknown

I've measured it from side to side:
'Tis three feet long, and two feet wide.

Wordsworth, William
Wordsworth Poetry and Prose
The Thorn
iii (Early Reading)

MODELS

A theory has only the alternative of being right or wrong. A model has a third possibility: it may be right, but irrelevant.

Eigen, Manfred
The Physicist's Conception of Nature
edited by Jagdish Mehra (p. 618)

Models are often used to decide issues in situations marked by uncertainty. However statistical differences from data depend on assumptions about the process which generated these data. If the assumptions do not hold, the inferences may not be reliable either. This limitation is often ignored by applied workers who fail to identify crucial assumptions or subject them to any kind of empirical testing. In such circumstances, using statistical procedures may only compound the uncertainty . . . Statistical modeling seems likely to increase the stock of things you think you know that ain't so.

Greedman, D.A.
Navidi, W.C.
Statistical Science
Regression Models for Adjusting the 1980 Census
Volume 1, Number 1, 1986 (p. 3)

The words "model" and "mode" have, indeed, the same root; today, model building is science *à la mode*.

Kaplan, Abraham
The Conduct of Inquiry
Chapter VII, Section 30 (p. 258)

The purpose of models is not to fit the data but to sharpen the questions.

Karlin, Samuel
11th R.A. Fisher Memorial Lectures
Royal Society 20 April 1983

Nay, Knowledge must come through action; thou canst have no test which is not fanciful, save by trial.

Sophocles
The Plays of Sophocles
Trachiniae
l. 589

The sciences do not try to explain, they hardly even try to interpret, they mainly make models.

Unknown

BET YOU HE'S A STATISTICIAN!

(Statistician) A figure head
Evan Esar – (See p. 223)

OBSERVATIONS

No observations are absolutely trustworthy.

Anscombe, F.J.
Technometrics
Rejection of Outliers
Volume 2, 1960 (p. 124)

. . . while those whom devotion to abstract discussions has rendered unobservant of the facts are too ready to dogmatize on the basis of a few observations.

Aristotle
On Generation and Corruption
Book I, Chapter II

Consider that everything which happens, happens justly, and if thou observest carefully, thou wilt find it to be so.

Aurelius, Marcus
The Meditations of the Emperor Antonius Marcus Aurelius
Book IV, Section 10

Speaking concretely, when we say "making experiments or making observations," we mean that we devote ourselves to investigation and to research, that we make attempts and trials in order to gain facts from which the mind, through reasoning, may draw knowledge or instruction.

Speaking in the abstract, when we say, "relying on observation and gaining experience," we mean that observation is the mind's support in reasoning, and experience the mind's support in deciding, or still better, the fruit of exact reasoning applied to the interpretation of facts.

Observation, then, is what shows facts; experiment is what teaches about facts and gives experience in relation to anything.

Bernard, Claude
An Introduction to the Study of Experimental Medicine (p. 11)

You can observe a lot by just watching.

Berra, Yogi
Quoted in Dick Schaap and Mort Gerberg's
Joy in Mudville: The Big Book of Baseball Humor
Reflections (p. 185)

A fool sees not the same tree that a wise man sees.

Blake, William
The Complete Writings of William Blake
The Marriage of Heaven and Hell
Proverbs of Hell
l. 8

To find out what happens to a system when you interfere with it you have to interfere with it (not just passively observe it).

Box, G.E.P.
Technometrics
Use and Abuse of Regression
Volume 8, Number 4, November 1966 (p. 629)

Shakespeare says, we are creatures that look before and after: the more surprising that we do not look round a little, and see what is passing under our very eyes.

Carlyle, Thomas
Sartor Resartus
Book I, Chapter 1

Oh, he is a good observer, but he has no power of reasoning!

Darwin, Charles
The Life and Letters of Charles Darwin
Volume I
Mental Qualities (p. 82)

For no one is so weak in mind that he does not perceive that while he is seated he is in some way different from what he is when he is standing on his feet.

Descartes, René
Rules for the Direction of the Mind
Rule XII

The bearing of this observation lays in the application on it.

Dickens, Charles
The Work of Charles Dickens
Dombey and Son
Chapter 23

A man should look for what is, and not for what he thinks should be . . .

Einstein, Albert
Quoted in Peter Michelmore's
Einstein (p. 20)

Ettore's Observation: The Other Line moves faster. This applies to all lines—bank, supermarket, tollbooth, customs, and so on. And don't try to change lines. The Other Line—the one you were in originally—will then move faster.

Ettore, Barbara
Harper's Magazine
Volume 249, Number 1491, August 1974

You must acquire the ability to describe your observations and your experience in such language that whoever observes or experiences similarly will be forced to the same conclusion.

Fabing, Harold
Mar, Ray
Fischerisms (p. 8)

. . . the link between observation and formulation is one of the most difficult and crucial in the scientific enterprise. It is the process of interpreting our theory or, as some say, of "operationalizing our concepts". Our creations in the world of possibility must be fitted in the world of probability; in Kant's epigram, "Concepts without precepts are empty". It is also the process of relating our observations to theory; to finish the epigram, "Precepts without concepts are blind".

Greer, Scott
The Logic of Social Inquiry (p. 160)

This assumption is not permissible in atomic physics; the interaction between observer and object causes uncontrollable and large changes in the system being observed, because of the discontinuous changes characteristic of atomic processes.

Heisenberg, W.
The Physical Principles of the Quantum Theory
Introductory (p. 3)

You see, but you do not observe. The distinction is clear.

Holmes, Sherlock
in Arthur Conan Doyle's
The Complete Sherlock Holmes
A Scandal in Bohemia

Never trust impressions, my boy, but concentrate yourself upon details.

> **Holmes, Sherlock**
> in Arthur Conan Doyle's
> *The Complete Sherlock Holmes*
> A Case of Identity

The way into my parlor is up a winding stair,
And I have many curious things to show when you are there . . .

> **Howitt, Mary**
> *The Poems of Mary Howitt*
> The Spider and the Fly

Seeing many things, but thou observest not . . .

> *The Bible*
> Isaiah 42:20

I do love to note and to observe.

> **Jonson, Ben**
> *Volpone*
> Act II, Scene 1

Although by now a large amount of observational material is available, the implications of the observations are far from clear.

> **Longair, M.S.**
> *Contemporary Physics*
> Quasi-Stellar Radio Sources
> Volume 8, 1967

But I keep no log of my daily grog,
For what's the use o' being bothered?
I drink a little more when the wind's offshore,
And most when the wind's from the no'th'ard.

> **Macy, Arthur**
> *Poems*
> The Indifferent Mariner

It urges the scientist, in effect, not to take risks incurred in moving far from the facts. However, it may properly be asked whether science can be undertaken without taking the risk of skating on the possibly thin ice of supposition. The important thing to know is when one is on the more solid ground of observation and when one is on the ice.

> **O'Neil, W.M.**
> *Fact and Theory*
> Chapter 8 (p. 154)

. . . to observe is not enough. We must use our observations, and to do that we must generalize.

Poincaré, Henri
The Foundations of Science
Science and Hypothesis (p. 127)

To observations which ourselves we make,
We grow more partial for th' observer's sake.

Pope, Alexander
The Complete Poetical Works of POPE
Moral Essays
Epis. I, l. 11

Some scientists find, or so it seems, that they get their best ideas when smoking; others by drinking coffee or whiskey. Thus there is no reason why I should not admit that some may get their ideas by observing or by repeating observations.

Popper, Karl R.
Realism and the Aim of Science (p. 36)

Keine Antwort ist auch eine Antwort.
[No answer is also an answer.]

Proverb, German

De lo que veas, cree muy poco,
De lo que te cuenten, nada.
[Of what you see, believe very little,
Of what you are told, nothing.]

Proverb, Spanish

I will tell you a moment in my life when I almost missed learning something. It was during the war and I was a farm laborer and my task was before breakfast to go to yonder hill and to a field there and count the cattle. I went and I counted the cattle—there were always thirty-two—and then I went back to the bailiff, touched my cap, and said, "Thirty-two, sir." and went and had my breakfast. One day when I arrived at the field an old farmer was standing at the gate, and he said, "Young man, what do you do here every morning?" I said, "Nothing much. I just count the cattle." He shook his head and said, "If you count them every day they won't flourish." I went back, I reported thirty-two, and on the way back I thought, Well, after all, I am a professional statistician, this is only a country yokel, how stupid can he get. One day I went back, I counted and counted again, there were only thirty-one. The bailiff was very angry. He said, "Have your breakfast and then we'll go up there together." And we went together and we searched the place and indeed, under a bush, was a dead beast. I thought to myself, Why have I been counting them all the time? I haven't prevented this beast dying. Perhaps that's what the farmer meant. They won't flourish if you don't look and watch the quality of each individual beast. Look him in the eye. Study the sheen on his coat. Then I might have gone back and said, "Well, I don't know how many I saw but one looks mimsey."

Schumacher, E.F.
Good Work
Education for Good Work (p. 145)

You will see something new.
Two things. And I call them
Thing One and Thing Two.

Seuss, Dr.
The Cat in the Hat (p. 33)

The observed of all observers . . .

Shakespeare, William
The Complete Works of William Shakespeare
Hamlet, Prince of Denmark
Act III, Scene 1, l. 162

. . . and in his brain, . . . he hath strange places cramm'd with observations . . .

Shakespeare, William
The Complete Works of William Shakespeare
As You Like It
Act II, Scene 7, l. 38

That was excellently observ'd, say I, when I read a Passage in an Author, where his Opinion agrees with mine. When we differ, there I pronounce him to be mistaken.

Swift, Jonathan
Satires and Personal Writings
Thoughts on Various Subjects

'Tis here, 'tis there. 'Tis gone.

Whitehead, Alfred North
An Introduction to Mathematics (p. 1)

ORDER

. . . altho' Chance produces Irregularities, still the Odds will be infinitely great, that in the process of Time, those Irregularities will bear no proportion to the recurrency of that Order which naturally results from ORIGINAL DESIGN *. . .* such Laws, as well as the original Design and Purpose of their Establishment, must all be *from without . . .* if we blind not ourselves with metaphysical dust, we shall be led, by a short obvious way, to the acknowledgment of the great MAKER and GOVERNOUR of all; *Himself all-wise, all-powerful* and *good.*

de Moivre, Abraham
The Doctrine of Chances (pp. 251–2)

The order is rapidly fading
And the first one now will later be last.

Dylan, Bob
"The Times They Are A-Changin' "

If you take a pack of cards as it comes from the maker and shuffle it for a few minutes, all trace of the original systematic order disappears. The order will never come back however long you shuffle. Something has been done which cannot be undone, namely, the introduction of a random element in place of the arrangement.

Eddington, Sir Arthur Stanley
The Nature of the Physical World (p. 63)

For in very truth, not by design did the first-beginnings of things place themselves each in their order with foreseeing mind, nor indeed did they make compact what movements each should start, but because many of them shifting in many ways throughout the world are harried and buffeted by blows from limitless time, by trying movements and unions of every kind, at last they fall into such dispositions as those, whereby our world of things is created and holds together.

Lucretius
Lucretius On the Nature of Things
Book I, 1020

Order is heaven's first law.

Pope, Alexander
The Complete Poetical Works of POPE
An Essay on Man
Epistle IV, l. 49

No answer is also an answer.
German Proverb – (See p. 146)

OUTLIERS

The fact that something is far-fetched is no reason why it should not be true; it cannot be as far-fetched as the fact that something exists.

Green, Celia
The Decline and Fall of Science
Aphorisms (p. 1)

I don't see the logic of rejecting data just because they seem incredible.

Hoyle, Fred
Quoted in D.O. Edge and M.J. Mulkay's
Astronomy Transformed (p. 432)

In almost every true series of observations, some are found, which differ so much from the others as to indicate some abnormal source of error not contemplated in the theoretical discussions, and the introduction of which into the investigations can only serve, in the present state of science, to perplex and mislead the inquirer.

Peirce, Benjamin
The Astronomical Journal (p. 160)

The folly of rejecting an extreme observation was demonstrated when shortly after 7 AM on the morning of December 7, 1941, the officer in charge of a Hawaiian radar station ignored data solely because it seemed so incredible.

Unknown

PERCENTAGES

There's a 50 percent chance of anything—either it happens or it doesn't.

Barnes, Michael R.
Quoted in Paul Dickson's
The Official Explanations (p. B-9)

Ninety per cent of everything is crap

Bloch, Arthur
Murphy's Law
Sturgeon's Law

"I did," Gerhard said. "But I don't know any more. We've passed the confidence limits already. They were about plus or minus two minutes for ninety-nine percent."

Crichton, Michael
The Terminal Man
Chapter 6 (p. 157)

John. Trust us on this, we have the figures. We are telling you with ninety-five percent confidence intervals how the people feel.

Crichton, Michael
Rising Sun
Second Day (p. 255)

My eye was caught this morning by a statement in the paper that "76 percent of adults have bad breath". I am always puzzled by such dogmatic observations. How are these conclusions reached? Do investigators scamper about the streets, sniffing?

Davies, Robertson
The Diary of Samuel Marchbanks
Winter, Section IV, Wednesday (p. 17)

Using any reasonable definition of a scientist, we can say that between 80 and 90 percent of all the scientists that have ever lived are alive now.

de Solla Price, Derek John
Little Science, Big Science (p. 1)

Kissinger's concern about a Russian attack on China was expressed many times. I used to tease him about his use of percentages. He would say there was a 60 percent chance of a Soviet strike on China for example, and I would say, "Why 60, Henry? Couldn't it be 65 percent or 58 percent?"

Haldeman, H.R.
The Ends of Power
Book Three (p. 89)

It's a little like the tale of the roadside merchant who was asked to explain how he could sell rabbit sandwiches so cheap. "Well" he explained, "I have to put in some horse meat too. But I mix them 50–50. One horse, one rabbit."

Huff, Darrell
How to Lie with Statistics (p. 114)

When half a million babies are born in England in a year, we may say that 20 percent of them are born in London, 2 percent in Manchester, 1 percent in Bristol, and so on. But when we think of one baby born in a single minute of time, we cannot say that 20 percent of it was born in London, 2 percent in Manchester, and so on. We can only say that there is a 20 percent probability of its being born in London, a 2 percent probability of its being born in Manchester, and so on.

Jeans, James Hopwood
Physics and Philosophy
Chapter V (p. 136)

When the weather predicts 30 percent chance of rain, rain is twice as likely as when 60 percent chance is predicted.

Parry, Thomas
Quoted in Paul Dickson's
The Official Explanations (p. P-175)

Later that evening we were watching the news, and the TV weathercaster announced that there was a 50 percent chance of rain for Saturday, and a 50 percent chance for Sunday, and concluded that there was therefore a 100 percent chance of rain that weekend.

Paulos, John Allen
Innumeracy (p. 3)

"That would be a little like saying '102 percent normal,'" said the Master smugly.

"If you like statistical scales better than the truth," Bux growled.

Sturgeon, Theodore
Quoted in Harlon Ellison's
Dangerous Visions
If All Men Were Brothers, Would You Let One Marry Your Sister? (p. 350)

. . . I do not remember just when, for I was not born then and cared nothing for such things. It was a long journey in those days and must have been a rough and tiresome one. The village contained a hundred people and I increased the population by 1 percent. It was more than many of the best men in history could have done for a town. It may not be modest in me to refer to this but it is true.

Twain, Mark
The Autobiography of Mark Twain
Chapter 1

"Well", he explained,
"I have to put in some horse meat too. But I mix them 50-50.
One horse, one rabbit."
Darrell Huff –
(See p. 152)

PRAYER

Thank God for compensating errors.

Fiedler, Edgar R.
Across the Board
The Three R's of Economic Forecasting—Irrational, Irrelevant and Irreverent
June 1977

Lord, please find me a one-armed statistician . . . so I won't always hear 'on the other hand . . .'

Hammond, Kenneth R.
Adelman, Leonard
Paraphrasing Edmund Muskie
Science
Science, Values, and Human Judgment
Volume 194, Number 4263, 22 October 1976 (p. 390)

What is the thing we call Common Sense? It is prayer practically applied; assistance given hope.

Howe, E.W.
Sinner Sermons (p. 7)

I call upon God, and beg him to be our savior out of a strange and unwanted enquiry, and to bring us to the heaven of probability.

Plato
Timaeus
48

The physical sciences are used to "praying over" their data, examining the same data from a variety of points of view. This process has been very rewarding, and has led to many extremely valuable insights. Without this sort of flexibility, progress in physical science would have been much slower. Flexibility in analysis is often to be had honestly at the price of a willingness not to demand that what has *already* been observed shall establish, or prove, what analysis *suggests*. In physical science generally, the results of praying over the data are thought of as something to be put to further test in another experiment, as indications rather than conclusions.

Tukey, John W.
The Annals of Mathematical Statistics
The Future of Data Analysis
Volume 33, Number 1, March 1962 (p. 46)

PREDICTION

The aim of every science is foresight (*prevoyance*). For the laws of established observation of phenomena are generally employed to foresee their succession. All men, however little advanced make true predictions, which are always based on the same principle, the knowledge of the future from the past.

Compte, Auguste
Quoted in Bertrand de Jouvenel's
The Art of Conjecture (p. 111)

Cutting up fowl to predict the future is, if done honestly and with as little interpretation as possible a kind of randomization. But chicken guts are hard to read and invite flights of fancy or corruption.

Hacking, Ian
The Emergence of Probability
Am Absent Family of Ideas (p. 3)

. . . if we can predict successfully on the basis of a certain explanation we have good reason, and perhaps the best of reason, for accepting the explanation.

Kaplan, Abraham
The Conduct of Inquiry
Chapter IX, Section 40 (p. 350)

Wall Street indexes predicted nine out of the last five recessions!

Samuelson, Paul A.
Newsweek
Science and Stocks
September 19, 1966 (p. 92)

"Hold your peace, old soothsayer," said Heriot, who at that instant entered the room with a calm and steady countenance. "Your calculations are true and undeniable when they regard brass and wire and mechanical force; but future events are at the pleasure of Him who bears the hearts of kings in His hands."

Scott, Sir Walter
The Fortunes of Nigel
Chapter VI (p. 75)

To predict is one thing. To predict correctly is another.

Unknown

The most we can know is in terms of probabilities.
Richard P. Feynman – (See p. 167)

PROBABILITY

FORD: Arthur, This is fantastic, we've been picked up by a ship with the new Infinite Improbability Drive, this is really incredible, Arthur . . . Arthur, what's happening?

ARTHUR: Ford, there's an infinite number of monkeys outside who want to talk to us about this script for Hamlet they've worked out.

Adams, Douglas
The Original Hitchhiker's Guide to the Galaxy Radio Script
Fit the Second (pp. 41–2)

TRILLIAN: Five to one against and falling . . . four to one against and falling . . . three to one . . . two . . . one . . . Probability factor one to one . . . we have normality . . . I repeat we have normality . . . anything you still can't cope with is therefore your own problem.

Adams, Douglas
The Original Hitchhiker's Guide to the Galaxy Radio Script
Fit the Second (p. 42)

The Reader may here observe the Force of Numbers, which can be successfully applied, even to those things, which one would imagine are subject to no Rules. There are very few things which we know, which are not capable of being reduc'd to a Mathematical Reasoning, and when they cannot, its a Sign our Knowledge of them is very small and confus'd; And where mathematical reasoning can be had, its as great folly to make use of any other, as to grope for a thing in the dark, when you have a Candle standing by you. I believe the Calculation of the Quantity of Probability might be improved to a very useful and pleasant Speculation, and applied to a great many Events which are accidental, besides those of Games; . . .

Arbuthnot, John
Of the Laws of Chance
Preface

The calculus of probabilities, when confined within just limits, ought to interest, in an equal degree, the mathematician, the experimentalist, and the statesman. From the time when Pascal and Fermat established its first principles, it has rendered, and continues daily to render, services of the most eminent kind. It is the calculus of probabilities, which, after having suggested the best arrangements of the tables of populations and mortality, teaches us to deduce from those numbers, in useful character; it is the calculus of probabilities which alone can regulate justly the premiums to be paid for assurances; the reserve funds for the disbursements of pensions, annuities, discounts, etc. It is under its influence that lotteries and other shameful snares cunningly laid for avarice and ignorance have definitely disappeared.

Arago
Smithsonian Report
Eulogy on Laplace
1874 (p. 164)

For that which is probable is that which generally happens.

Aristotle
The Art of Rhetoric
Book I, Chapter II

The good or evil of an event should be considered in view of the event's likelihood of occurrence.

Arnauld, Antoine
The Art of Thinking: Port-Royal Logic
Belief in future contingent events (p. 355)

Probability and Birds in the Yard

Atkins, Russell
Title of poem

Are no probabilities to be accepted, merely because they are not certainties?

Austen, Jane
Sense and Sensibility
Volume I, Chapter 15

Life is a school of probability.

Bagehot, Walter
Quoted in Rudolf Flesch's
The New Book of Unusual Quotations

The more ridiculous a belief system, the higher the probability of its success.

Bartz, Wayne R.
Human Behavior
Keys to Success

Luck was not probability, but it acted through probability. It was, so to speak, quantities of probability, a quantitative average throughout the universe. And like any other fixed quantity, it could only be concentrated or increased at the cost of a diminution elsewhere.

Bayley, Barrington J.
The Grand Wheel (p. 151)

Ambrozial weather will permeate all around Pordunk for the next 16 months, with rain and snow, and all sorts ov stuff in the ballance ov the United States of America.

The probabilitiz that the abuv probabilitiz will assimilate themselfs tew the principal probabilitiz in the case.

If they don't, du notiss will be giv.

In the mean time be kalm, be dignified, and don't be skeerd.

Billings, Josh
Old Probability: Perhaps Rain—Perhaps Not
Probabilitiz 1873

. . . all is to them a dull round of probabilities and possibilities.

Blake, William
The Complete Writings of William Blake
The Ancient Britons

Probability is expectation founded upon partial knowledge. A perfect acquaintance with *all* the circumstances affecting the occurrence of an event would change expectation into certainty, and leave neither room nor demand for a theory of probabilities.

Boole, George
Collected Logical Works
Volume II
An Investigation of the Law of Thought
Chapter XVI (p. 258)

Probabilities must be regarded as analogous to the measurement of physical magnitudes; that is to say, they can never be known exactly, but only within certain approximation.

Borel, Emile
Probabilities and Life
Introduction (pp. 32–3)

It is easier to make true misleading statements in the subject of probabilities than anywhere else.

Bostwick, Arthur E.
Science
The Theory of Probabilities
Volume III, Number 54, January 10, 1896 (p. 66)

Johnson. "If I am well acquainted with a man, I can judge with great probability how he will act in any case, without his being restrained by my judging. God may have this probability increased to certainty."

Boswell, James
The Life of Samuel Johnson
Volume II
April 15, 1778 (pp. 209–10)

Probability tells us what we ought to believe, what we ought to believe *on certain data* . . . Probability is no more 'relative' and 'subjective' than is any other act of logical inference from hypothetical premise.

Bradley, F.H.
The Principles of Logic (p. 208)

Fate laughs at probabilities.

Bulwer, Lytton, E.G.
Eugene Aram
Book I
Chapter 10 (p. 71)

The play of imagination, in the romance of early youth, is rarely interrupted with scruples of probability.

Burney, Fanny
Camilla
Book II, Chapter V (p. 102)

But to *us*, probability is the very guide to life.
Butler, Joseph
The Analogy of Religion
Introduction (p. xxv)

. . . law, like other branches of social science, must be satisfied to test the validity of its conclusions by the logic of probabilities rather than the logic of certainty.

Cardozo, Benjamin N.
The Growth of the Law (p. 33)

"Do you think that to cut a man's throat like that would need a great force? Or, perhaps, only a very sharp knife?"

"I should say that it could not be done with a knife at all," said the pale doctor.

"Have you any thought," resumed Valentine, "of a tool with which it could be done?"

"Speaking within modern probabilities, I really haven't," said the doctor, arching his painful brow.

Chesterson, Gilbert Keith
The Father Brown Omnibus
The Innocence of Father Brown
The Secret Garden

. . . the electron is just a "smear of probability".

Coats, R.H.
Journal of the American Statistical Association
Science and Society
Volume 34, Number 205, March 1939 (p. 6)

Unlike almost all mathematics, I agree completely with your statement that every probability evaluation is a probability evaluation, that is, something to which it is meaningless to apply such attributes as *right*, *wrong*, *rational*, etc.

Cohen, John
Chance, Skill, and Luck
Chapter 2, Part 1 (p. 28)

Considine's Law. Whenever one word or letter can change the entire meaning of a sentence, the probability of an error being made will be in direct proportion to the embarrassment it will cause.

Considine, Bob
Quoted in Paul Dickson's
The Official Rules (p. C-32)

Harry sighed irritably, pulled out a sheet of paper. "It's a probability equation." He wrote:

$$p = f_p n_h f_l f_i f_c$$

"What it means," Harry Adams said, "is that the probability, p, that intelligent life will evolve in any star system is a function of the probability that the star will have planets, the number of habitable planets, the probability that simple life will evolve on a habitable planet, the probability that intelligent life will evolve from simple life, and the probability that intelligent life will attempt interstellar communication within five billion years. That's all the equation says."

Crichton, Michael
Sphere
The Briefing (p. 29)

"But the point is that we have no facts," Harry said. "We must guess at every single one of these probabilities."

Crichton, Michael
Sphere
The Briefing (p. 29)

The mathematical theory of probability is a science which aims at reducing to calculation, where possible, the amount of credence due to propositions or statements, or to the occurrence of events, future or past, more especially as contingent or dependent upon other propositions or events the probability of which is known.

Crofton, M.W.
The Encyclopaedia Britannica
9th Edition
Probability

Indeed the intellectual basis of all empirical knowledge may be said to be a matter of probability, expressible only in terms of a bet.

Dampier-Whetham, William
A History of Science
Chapter III (p. 155)

As for a future life, every man must judge for himself between conflicting vague probabilities.

Darwin, Charles
The Life and Letters of Charles Darwin
Religion (p. 277)

. . . I should reply that the falsehood is all the greater when it appears in the guise of truth, and that as fiction, the more it contains of the pleasing and the possible the more it delights us.

de Cervantes, Miguel
The Ingenious Gentleman Don Quixote de la Mancha
Part I, Chapter 47

My thesis . . . is simply this:

PROBABILITY DOES NOT EXIST.

The abandonment of superstitious beliefs about . . . Fairies and Witches was an essential step along the road to scientific thinking. Probability, too, if regarded as something endowed with some kind of objective existence, is no less a misleading conception, an illusory attempt to exteriorize or materialize our true probabilistic beliefs.

In investigating the reasonableness of our own modes of thought and behavior under uncertainty, all we require, and all that we are reasonably entitled to, is consistency among these beliefs, and their reasonable relation to any kind of relevant objective data . . . This is Probability Theory.

de Finetti, B.
Theory of Probability (p. x)

We defined the art of conjecture, or stochastic art, as the art of evaluating as exactly as possible the probabilities of things, so that in our judgments and actions we can always base ourselves on what has been found to be the best, the most appropriate, the most certain, the best advised; this is the only object of the wisdom of the philosopher and the prudence of the statesman.

de Jouvenel, Bertrand
The Art of Conjecture
(p. 21, note 19)

But how is it they suffer themselves to incline to and be swayed by probability, if they know not the truth itself?

de Montaigne, Michel Eyquem
The Essays
Essays II, 12

Deifield's Principle. The probability of a young man meeting a desirable receptive young female increases by pyrimidical progression when he is already in the company of (1) a date, (2) his wife, (3) a better looking and richer male friend.

Deifield, Ronald H.
Quoted in Paul Dickson's
The Official Rules (p. B-12)

The statistician's report to management should not talk about probabilities. It will merely give outside margins of error for the results of chief importance.

Deming, William Edwards
Sample Design in Business Research (p. 13)

As to the influence and genius of great generals—there is a story that Enrico Fermi once asked Gen. Leslie Groves how many generals might be called "great". Groves said about three out of every 100. Fermi asked how a general qualified for the adjective, and Groves replied that any general who had won five major battles in a row might safely be called great. This was in the middle of World War II. Well, then, said Fermi, considering that the opposing forces in most theaters of operation are roughly equal, the odds are one of two that a general will win a battle, one of four that he will win two battles in a row, one in eight for three, one of sixteen for four, one of thirty-two for five. "So you are right, General, about three out of every 100. Mathematical probability, not genius."

Deming, William Edwards
Out of the Crisis (p. 394)

There can be no unique probability attached to any event or behaviour: we can only speak of "probability in the light of certain given information", and the probability alters according to the extent of the information.

Eddington, Sir Arthur Stanley
The Nature of the Physical World (p. 315)

One difficulty in employing strength of belief as a measure of probability is that an expectation of belief has partly a subjective bias. We have agreed that it depends (and ought to depend) on the information or evidence supplied; but in addition the strength of the expectation depends on the personality of the man who weighs the evidence. We try to remove this subjective element by saying that the true probability corresponds to the judgment of a 'right-thinking person'; but how shall we define this ideal reference? We do not mean a perfectly logical person, for there is no question of making a strictly logical deduction from the evidence; if that were possible the conclusion would be a matter of certainty not probability. We do not mean a person gifted with second-sight, for we want to know the probability relative to the information stated and not relative to occult information. We do not particularly mean a person of long experience in similar judgments, for he is likely to use his past experience to supplement surreptitiously the information on which the judgment of probability is ostensibly based. Apart from the obvious definition of a right-thinking person as 'someone who thinks as I do' (which is probably the definition at the back of our minds) there seems to be no easy way of defining his qualities.

Eddington, Sir Arthur Stanley
New Pathways in Science
Probability (p. 112)

In most modern theories of physics probability seems to have replaced aether as "the nominative of the verb 'to undulate'".

Eddington, Sir Arthur Stanley
New Pathways in Science
Probability (p. 110)

Probability may be described, agreeably to general usage, as importing partial incomplete belief.

Edgeworth, Francis Ysidro
Mind
The Philosophy of Chance
Volume 9, 1884

. . . ignorance gives one a large range of probabilities.

Eliot, George
Daniel Deronda II
xiii (p. 100)

Secrets are rarely betrayed or discovered according to any program our fear has sketched out. Fear is almost always haunted by terrible dramatic scenes, which recur in spite of the best-argued probabilities against them . . .

Eliot, George
The Mill on the Floss
Book V, V

Still there is a possibility—even a probability—the other way.

Eliot, George
The George Eliot Letters
Volume II (p. 127)

But I see no probability of my being able to be with you before your other Midsummer visitors arrive.

Eliot, George
The George Eliot Letters
Volume II (p. 160)

Fourth Law of Thermodynamics. If the probability of success is almost one, then it is damn near zero.

Ellis, David
Quoted in Paul Dickson's
The Official Rules (p. F-60)

Probability is a mathematical discipline with aims akin to those, for example, of geometry or analytical mechanics. In each field we must carefully distinguish three aspects of the theory: (a) the formal logical content, (b) the intuitive background, (c) the applications. The character, and the charm, of the whole structure cannot be appreciated without considering all three aspects in their proper relation.

Feller, William
An Introduction to Probability Theory and Its Applications (p. 1)

All possible "definitions" of probability fall short of the actual practice.

Feller, William
An Introduction to Probability Theory and Its Applications (p. 19)

In its efforts to learn as much as possible about nature, modern physics has found that certain things can never be "known" with certainty. Much of our knowledge must always remain uncertain. The *most* we can know is in terms of probabilities.

Feynman, Richard P.
The Feynman Lectures on Physics (pp. 6–11)

. . . the ratios or probabilities of which we have been speaking have no *absolute* signification with reference to an event which *has* occurred . . . They represent only the state of *expectation* of the mind of a person before the event has occurred, or having occurred before he is informed of the results.

Forbes, J.D.
The London, Edinburgh and Dublin Philosophical Magazine and Journal of Science
On the alleged evidence for a Physical Connection between
Stars forming Binary or Multiple Groups
Third Series, December 1850 (p. 406)

It is a question of probabilities . . .

Freeman, R. Austin
A Certain Dr. Thorndyke
Thorndyke Makes a Beginning

The balance of probabilities is in favor of that view.

Friedman, Thomas L.
From Beirut to Jerusalem (p. 35)

Philosophy goes no further than probabilities, and in every assertion keeps a doubt in reserve.

Froude, James Anthony
Short Studies on Great Subjects
Calvinism (p. 51)

After all, without the experiment—either a real one or a mathematical model—there would be no reason for a theory of probability.

Fry, Thornton C.
Probability and Its Engineering Uses (p. 15)

But if probability measures the importance of our state of ignorance it must change its value whenever we add new knowledge. And so it does.

Fry, Thornton C.
Probability and Its Engineering Uses (p. 145)

Lest men suspect your tale untrue,
Keep probability in view.

Gay, John
John Gay: Poetry and Prose
Fables
The Painter Who Pleased Nobody and Everybody
l. 1

Men are deplorably ignorant with respect to natural things, and modern philosophers, as though dreaming in the darkness, must be aroused and taught the uses of things, the dealing with things; they must be made to quit the sort of learning that comes only from books, and that rests only on vain arguments from probability and upon conjecture.

Gilbert, William
On the Loadstone and Magnetic Bodies and on the Great Magnet the Earth
Book 1, Chapter 10

Of course, if your work is strong, and you can afford to wait, the probability is that half a dozen people will at last begin to shout that you have been monstrously neglected, as you have.

Gissing, George
New Grub Street
Interim (p. 411)

It is only by mature meditation on the possibilities and probabilities of future events—that we can elude the tortuous troubles of the tomorrows.

Gracian, Balthasar
Quoted in Thomas G. Corvan's
The Best of Gracian (p. 22)

Whereas wisdom favors the probabilities, folly favors only the possibilities.

Gracian, Balthasar
Quoted in Thomas G. Corvan's
The Best of Gracian (p. 38)

Wisdom does not trust to probabilities; it always marches in the midday light of reason.

Gracian, Balthasar
Quoted in Rudolf Flesch's
The New Book of Unusual Quotations

... the contradictory of a welcome probability will assert itself whenever such an eventuality is likely to be most frustrating.

Gumperson, R.F.
Changing Times
Gumperson's Law
Volume 11, Number 11, November, 1957 (p. 46)

The outcome of a given desired probability will be inverse to the degree of desirability.

Gumperson, R.F.
Changing Times
Gumperson's Law
Volume 11, Number 11, November, 1957 (p. 46)

Probability is too important to be left to the experts.

Hamming, Richard W.
The Art of Probability for Scientists and Engineers (p. 4)

Probability is truth in some degree.

Harris, Errol E.
Hypothesis and Perception
The Logic of Construction (p. 342)

probability = (possibility)^2

Herbert, Nick
Quantum Reality (p. 96)

No priest or soothsayer that ever lived could hold his own against Old Probabilities.

Holmes, O.W.
Pages from an Old Volume of Life (p. 327)

As for probabilities, what thing was there ever set down so agreeable with sound reason but some probable show against it might be made.

Hooker, Richard
Quoted in S. Austin Allibone's
Prose Quotations from Socrates to Macaulay
Probability

A reasonable probability is the only certainty.

Howe, E.W.
Sinner Sermons (p. 23)

All knowledge resolves itself into probability.

Hume, David
A Treatise of Human Nature
Book I, Part IV, Section 1

"Now, your Honor; in much the same way that there are laws governing our society, there are also laws governing chance, and these are called the laws of probability, and it is against these that we must examine the use of an identical division number."

Hunter, Evan
The Paper Dragon
Tuesday
Chapter 6

Magic and devils offend our sense of probabilities.

Huxley, Aldous
Proper Studies
Varieties of Intelligence (p. 7)

. . . I have finally judged that it was better worth while to publish this writing such as it is, than to let it run the risk, by waiting longer, of remaining lost.

There will be seen in it demonstrations of those kinds which do not produce as great a certitude as those of geometry, and which even differ much therefrom, since, whereas the geometers prove their propositions by fixed and incontestable principles, here the principles are verified by the conclusions to be drawn from them; the nature of these things not allowing of this being done otherwise. It is always possible to attain thereby to a degree of probability which very often is scarcely less than complete proof. To wit, when things which have been demonstrated by the principles that have been assumed correspond perfectly to the phenomena which experiment has brought under observation; especially when there are a great number of them, and further, principally, when one can imagine and foresee new phenomena which ought to follow from the hypotheses which one employs, and when one finds that therein the fact corresponds to our prevision. But if all these proofs of probability are met with in that which I propose to discuss, as it seems to me they are, this ought to be a very strong confirmation of the success of my inquiry . . .

Huygens, Christiaan
Treatise on Light
Preface

"Juries hate scientific evidence."

"They think they won't be able to understand it so naturally they can't understand it. As soon as you step into the box you see a curtain of obstinate incomprehension clanging down over their minds. What they want is certainty. Did this paint particle come from this car body? Answer yes or no. None of those nasty mathematical probabilities we're so fond of."

James, P.D.
Death of an Expert Witness
Book II, Chapter III (p. 83)

Perhaps an editor might begin a reformation in some way as this. Divide his paper into four chapters, heading the 1st, Truth. 2nd, Probabilities. 3rd, Possibilities. 4th, Lies.

Jefferson, Thomas
Letter to John Norvell
June 11, 1807

To the author the main chain of probability theory lies in the enormous variability of its applications. Few mathematical disciplines have contributed to as wide a spectrum of subjects, a spectrum ranging from number theory to physics, and even fewer have penetrated so decisively the whole of our scientific thinking.

Kac, Mark
Lectures in Applied Mathematics
Volume I
Probability and Related Topics in Physical Sciences (p. ix)

Equiprobability in the physical world is purely a hypothesis . . . Thus, the statement "head and tail are equiprobable" is at best an assumption.

Kasner, Edward
Newman, James
Mathematics and the Imagination (p. 251)

. . . others have suggested seriously a 'barometer of probability'.

Keynes, John Maynard
A Treatise on Probability
Chapter III (p. 20)

Probability is, so far as measurement is concerned, closely analogous to similarity.

Keynes, John Maynard
A Treatise on Probability
Chapter III (p. 28)

It is difficult to find an intelligible account of the meaning of 'probability', or of how we are ever to determine the probability of any particular proposition; and yet treatises on the subject profess to arrive at complicated results of the greatest precision and the most profound practical importance.

Keynes, John Maynard
A Treatise on Probability
Chapter IV (p. 251)

The theory of probability as mathematical discipline can and should be developed from axioms in exactly the same way as Geometry and Algebra.

Kolmogorov, Andrei N.
Foundations of the Theory of Probability
Chapter 1
Elementary Theory of Probability (p. 1)

. . . there is no problem about probability: it is simply a non-negative, additive set function, whose maximum value is unity.

Kyburg, H.E., Jr. and Smokler, H.E.
Studies in Subjective Probability (p. 3)

It is remarkable that a science that began by considering games of chance should itself be raised to the rank of the most important subject of human knowledge.

Laplace, Pierre-Simon
A Philosophical Essay on Probabilities (p. 123)

The most important questions of life are, for the most part, really only problems of probability.

. . . in the small number of things we are able to know with any certainty . . . the principle means of arriving at the truth . . . are based on probabilities . . .

Laplace, Pierre-Simon
A Philosophical Essay on Probabilities (p. 1)

Probability has reference partly to our ignorance, partly to our knowledge.

Laplace, Pierre-Simon
Essai Philosophique sur les Probabilités (p. 9)

. . . the art of weighing probabilities is not yet even partly explained, though it would be of great importance in legal matters and even in the management business.

Leibniz, Gottfried Wilhelm
Leibniz: Philosophical Papers and Letters
Volume I
Letter to John Frederick, Duke of Brunswick Hanover (p. 399)

There is no such thing as *the* probability of four aces in one hand, or *the* probability of anything else. Given all the relevant data which there are to be known, everything is either certainly true or certainly false.

Lewis, Clarence Irving
Mind and the World-Order
Chapter X (p. 330)

A "poor evaluation" of the probability of anything may reflect ignorance of relevant data which "ought" to be known . . .

Lewis, Clarence Irving
Mind and the World-Order
Chapter X (p. 331)

. . . empirical knowledge is exclusively a knowledge of probabilities . . .

Lewis, Clarence Irving
Mind and the World-Order
Chapter XI (p. 345)

We may not be able to get certainty, but we can get probability, and half a loaf is better than no bread.

Lewis, C.S.
Christian Reflections
Para. 22 (p. 111)

The *probability* that we may fall in the struggle *ought not* to deter us from the support of a cause we believe to be just; it *shall not* deter me.

Lincoln, Abraham
The Sub-Treasury Speech
Springfield, Illinois
December 26, 1839

Are we probabilists, believers, or fuzzifiers?

Lindley, Dennis V.
Statistical Science
Comment: A Tale of Two Wells
Volume 2, Number 1
February 1987 (p. 38)

Probability is the appearance of agreement upon fallible proofs.

Locke, John
An Essay Concerning Human Understanding
Book IV, XV, 1

Probability is likeness to be true . . .

Locke, John
An Essay Concerning Human Understanding
Book IV, XV, 4

The mind ought to examine all the grounds of probability, and upon a due balancing the whole, reject or receive it proportionably to the preponderancy of probability on the one side or the other.

Locke, John
Quoted in S. Austin Allibone's
Prose Quotations from Socrates to Macaulay
Probability

It wasn't a probability anymore, it was a reality.

Ludlum, Robert
The Bourne Supremacy
Chapter 18 (p. 256)

It was a desperate strategy, based on probabilities, but it was all he had left.

Ludlum, Robert
The Bourne Supremacy
Chapter 24 (p. 365)

Messenger said, "Can you work out any equations of probability of one hitting here?"

"No sir. A hurricane has no memory. Like a coin. If a coin comes up heads fifty times, the odds on the next flip are still fifty–fifty, head or tail. But if you flip it ten thousand times, you'll get five thousand heads, plus or minus."

MacDonald, John D.
Condominium: A Novel
Chapter 26 (p. 235)

If absolutes had disappeared under the inquiries of science, and apparently they had, why then the only rational procedure, the only procedure consistent with man's development, was to follow where the probabilities led.

Masters, Dexter
The Accident (p. 19)

Uncertainty is introduced, however, by the impossibility of making generalizations, most of the time, which happens to all members of a class. Even scientific truth is a matter of probability and the degree of probability stops somewhere short of certainty.

Minnick, Wayne C.
The Art of Persuasion (p. 167)

The probability is, I suppose that the Monarchy has become a kind of ersatz religion. Chesterton once remarked that when people ceased to believe in God they do not believe in nothing, but in anything.

Muggeridge, Malcolm
New Statesman 1955

Take away *probability*, and you can no longer please the world; give *probability*, and you can no longer displease it.

Pascal, Blaise
The Thoughts of Blaise Pascal
Appendix: Polemical Fragments
918

Probabilities are summaries of knowledge that is left behind when information is transferred to a higher level of abstraction.

Pearl, Judea
Probabilistic Reasoning in Intelligent Systems (p. 21)

Hitherto the user has been accustomed to accept the function of probability theory laid down by the mathematicians; but it would be good if he could take a larger share in formulating himself what are the practical requirements that the theory should satisfy in applications.

Pearson, E.S.
Biometrika
The Choice of Statistical Test Illustrated on the Interpretation of
Data Classed in a 2×2 Table
Volume 34, Number 35, 1948 (p. 142)

. . . it may be doubtful if there is a single extensive treatise on probabilities in existence which does not contain solutions absolutely indefensible.

Peirce, Charles Sanders
Writings of Charles Sanders Peirce
Volume 3 (p. 278)

This branch of mathematics [probability] is the only one, I believe, in which good writers frequently get results entirely erroneous.

Peirce, Charles Sanders
Writings of Charles Sanders Peirce
Volume 3 (p. 279)

The idea of probability essentially belongs to a kind of inference which is repeated indefinitely. An individual inference must be either true or false, and can show no effect of probability; and, therefore, in reference to a single case considered in itself, probability can have no meaning.

Peirce, Charles Sanders
Writings of Charles Sanders Peirce
Volume 3 (p. 281)

I know too well that these arguments from probabilities are impostors, and unless great caution is observed in the use of them, they are apt to be deceptive.

Plato
Phaedo
92

"I think we can see the extent of the problem. I've measured harmonics up to the sixth-order already, and still propagating." He paused to look at the other faces for disagreement. There wasn't any. "If this goes on," he said evenly, "I project a nine-nines probability that within one standard year the disturbances will be effectively both plenary and irreversible."

Pohl, Frederik
The Coming of the Quantum Cats (p. 189)

The very name calculus of probabilities is a paradox. Probability opposed to certainty is what we do not know, and how can we calculate what we do not know?

Poincaré, Henri
The Foundations of Science
Science and Hypothesis (p. 155)

No matter how solidly founded a prediction may appear to us, we are never *absolutely* sure that experiment will not contradict it, if we undertake to verify it . . . It is far better to foresee even without certainty than not to foresee at all.

Poincaré, Henri
The Foundations of Science
Science and Hypothesis (p. 129)

The most important application of the theory of probability is to what we may call 'chance-like' or 'random' events, or occurrences. These seem to be characterized by a peculiar kind of incalculability which makes one disposed to believe—after many unsuccessful attempts—that all known rational methods of prediction must fail in their case. We have, as it were, the feeling that not a scientist but only a prophet could predict them. And yet, it is just this incalculability that makes us conclude that the calculus of probability can be applied to these events.

Popper, Karl R.
The Logic of Scientific Discovery (p. 150)

"You haven't heard of probability math? You, and tomorrow you become Chairman of the Board of Widdershins and heir to riches untold? Then first we will talk, and then we will eat."

Pratchett, T.
The Dark Side of the Sun (p. 13)

"I can't pretend to understand probability math. But if the universe is so ordered, so—*immutable*—that the future can be told by a handful of numbers, then why need we go on living?"

Pratchett, T.
The Dark Side of the Sun (p. 22)

"Understanding is the first step towards control. We now understand probability . . ."

Pratchett, T.
The Dark Side of the Sun (p. 37)

. . . by the mathemagic of probability, sifting through the population of the galaxy to find those who's probability profile matched the theoretical one for the discoverers of Jokers World.

Pratchett, T.
The Dark Side of the Sun (p. 153)

PROBABILITY MATH:

"As with the first Theory of Relativity and the Sadhimist One Commandment, so the nine equations of probability math provide an example of a deceptively simple spark initiating a great explosion of social change."

"Probability math predicts the future." So says the half-educated man . . .

"Probability math arises from the premise that we dwell in a truly infinite totality, space and time without limit, worlds without end—a creation so vast that what we are pleased to call our cause-and-effect datum Universe is a mere circle of candlelight. In such a totality we can only echo the words of Quixote: *All things are possible* . . ."

Pratchett, T.
The Dark Side of the Sun (p. 24)

In this case probability must atone for want of Truth.

Prior, Matthew
The Literary Works of Matthew Prior
Solomon
Preface (p. 309)

A thousand probabilities does not make one fact.

Proverb, Italian

I think I perceive or remember something but am not sure; this would seem to give me some ground for believing it, contrary to Mr. Keynes' theory, by which the degree of belief in it which it would be rational for me to have is that given by the probability relation between the proposition in question and the things I know for certain.

Ramsey, Frank Plumpton
The Foundation of Mathematics and Other Logical Essays
Truth and Probability
The Logic of Consistency (p. 190)

I feign probabilities. I record improbabilities.

Reade, Charles
A Terrible Temptation: a story of the day

Good and bad come mingled always. The long-time winner is the man who is not unreasonably discouraged by persistent streaks of ill fortune not at other times made reckless with the thought that he is fortune's darling. He keeps a cool head and trusts in the mathematics of probability, or as often said, the law of averages.

Redfield, Roy A.
Factors of Growth in a Law Practice (p. 168)

There is a special department of hell for students of probability. In this department there are many typewriters and many monkeys. Every time that a monkey walks on a typewriter, it types by chance one of Shakespeare's sonnets.

Russell, Bertrand A.
Nightmares of Eminent Persons
The Metaphysician's Nightmare (p. 29)

When we want something, we always have to reckon with probabilities.

Sartre, Jean-Paul
The Philosophy of Existentialism (p. 46)

It is better to be satisfied with probabilities than to demand impossibilities and starve.

Schiller, Friedrich
Quoted in Rudolf Flesch's
The New Book of Unusual Quotations

And nobody can get . . . far without at least an acquaintance with the mathematics of probability, not to the extent of making its calculations and filling examination papers with typical equations, but enough to know when they can be trusted, and when they are cooked. For when their imaginary numbers correspond to exact quantities of hard coins unalterably stamped with heads and tails, they are safe within certain limits; for here we have solid certainty . . . but when the calculation is one of no constant and several very capricious variables, guesswork, personal bias, and pecuniary interests, come in so strong that those who began by ignorantly imagining that statistics cannot lie end by imagining equally ignorantly, that they never do anything else.

Shaw, George Bernard
The World of Mathematics
Volume 3 (p. 1531)
The Vice of Gambling and the Virtue of Insurance

I hope that you flourish in Probabilities.

Letter from Francis Ysidro Edgeworth to Karl Pearson
11 September 1893
Quoted in Stephen M. Stigler's
The History of Statistics
Chapter 10 (p. 326)

If we postulate that within un-, sub- or supernatural forces *the probability is* that the law of probability will not operate as a factor, then we must accept that the probability of the *first* part will not operate as a factor with un-, sub- or supernatural forces. And since it obviously hasn't been doing so, we can take it that we are not held within un-, sub- or supernatural forces after all; in all probability, that is.

Stoppard, Tom
Rosencrantz and Guildenstern Are Dead
Act One (p. 17)

For we know in part, and we prophesy in part.

The Bible
I Corinthians 13:9

A pinch of probability is worth a pound of perhaps.

Thurber, James
Lanterns and Lances
Such a Phrase as Drifts Through Dreams

Though moral certainty be sometimes taken for a high degree of probability, which can only produce a doubtful assent, yet it is also frequently used for a firm assent to a thing upon such grounds as fully satisfy a prudent man.

Tollotson, John
Quoted in S. Austin Allibone's
Prose Quotations from Socrates to Macaulay
Probability

It is a known fact that if a man uses one of the end urinals his probability of being pissed on is reduced by 50 percent.

Unknown

He who has heard the thing told by twelve thousand eye-witnesses, has only twelve thousand probabilities, equal to one strong probability, which is not equal to certainty.

Voltaire
The Portable Voltaire
Philosophical Dictionary
Truth

From generation to generation skepticism increases; and probability diminishes; and soon probability is reduced to zero.

Voltaire
The Portable Voltaire
Philosophical Dictionary
Truth

Almost all human life depends on probabilities.

Voltaire
Essays
Probabilities

In short, absolute, so-called mathematical factors never find a firm basis in military calculations. From the very start there is an interplay of possibilities, probabilities, good luck and bad that weaves its way throughout the length and breadth of the tapestry. In the whole range of human activities war most closely resembles a game of cards.

von Clausewitz, Karl
On War
Chapter 1, 21 (p. 86)

The theory of probability can never lead to a definite statement concerning a single event.

von Mises, Richard
Probability, Statistics and Truth
First Lecture (p. 33)

. . . if one talks of the probability that the two poems known as the *Iliad* and the *Odyssey* have the same author, no reference to a prolonged sequence of cases is possible and it hardly makes sense to assign a *numerical* value to such a conjecture.

von Mises, Richard
Mathematical Theory of Probability and Statistics (pp. 13–4)

One can locate an octopus by giving the coordinates of his beak, but it would be unwise to forget that neighboring coordinates for two or three yards out in all directions have a considerable probability of being occupied by octopus at a given instant.

Walker, Marshall
The Nature of Scientific Thought (p. 65)

The road is a strange place. Shuffling along I looked up and you were there walking across the grass toward my truck on an August day. In retrospect, it seemed inevitable—it could not have been any other way—a case of what I call the high probability of the improbable.

Waller, Robert James
The Bridge of Madison County (pp. 22–3)

But Positivistic science is solely concerned with observed fact, and must hazard no conjecture as to the future. If observed fact be all we know, then there is no other knowledge. Probability is relative to knowledge. There is no probability as to the future within the doctrine of Positivism.

Whitehead, Alfred North
Adventures of Ideas
Cosmologies (p. 125)

Only a certain probability remains of a one-to-one association of any spatial feature *now* with a similar feature *a moment later*. It is sheer luck, in a sense, that any physical apparatus stays put, for the laws of quantum mechanics allow it a finite, though small, probability of dispersing while one is not looking, or even while one is.

Whyte, Lancelot Law
Essay on Atomism: from Democritus to 1960
Chapter 2 (pp. 25–6)

If the universe is a mingling of probability clouds spread through a cosmic eternity of space-time, how is there as much order, persistence, and coherent transformation as there is?

Whyte, Lancelot Law
Essay on Atomism: from Democritus to 1960
Chapter 2 (p. 27)

Gilbert . . . No ignoble consideration of probability, that cowardly concession to the tedious repetitions of domestic or public life, affect it ever.

Wilde, Oscar
The Critic as Artist
Part I

Ashley had no competitive sense and no need for money, but he took great interest in the play of numbers. He drew up charts analyzing the elements of probability in the various games. He had a memory for numbers and symbols.

Wilder, Thornton
The Eighth Day
II, Illinois to Chile (p. 123)

The theory of probabilities and the theory of errors now constitute a formidable body of knowledge of great mathematical interest and of great practical importance. Though developed largely through the applications to the more precise sciences of astronomy, geodesy, and physics, their range of applicability extends to all the sciences; and they are plainly destined to play an increasingly important role in the development and in the applications of the sciences of the future. Hence their study is not only a commendable element in a liberal education, but some knowledge of them is essential to a correct understanding of daily events.

Woodward, Robert S.
Probability and Theory of Errors
Preface

PROBABLE

In short, these fundamental elements of scientific knowledge assimilate and grow, coalesce and separate and recombine, shrink and wane, die and come to life again; and while they persist they are never more than probable.

Barry, Frederick
The Scientific Habit of Thought (p. 139)

. . . it is always probable that something improbable will happen.

Bleckley, Logan E.
Warren v. Purtell, 63 *Georgia Reports* 428, 430 (1879)

. . . many sensations are *probable*, that is, though not accounting to a full perception they are yet possessed of a certain distinctness and clearness, and can serve to direct the conduct of the wise man.

Cicero
De Natura Deorum
Book I, Chapter 5, section 12

The laws of chance tell us what is probable, but not what is certain to happen. They do not predict. They do not tell us what *will*, but what *may*, happen.

de Leeuw, A.L.
Rambling through Science
Gambling (p. 88)

When it is not in our power to determine what is true, we ought to act accordingly to what is most probable . . .

Descartes, René
Discourse on the Method of Rightly Conducting the Reason and Seeking for Truth in the Sciences
Part III

Such a fact is probable, but undoubtedly false.

Gibbon, Edward
The Decline and Fall of the Roman Empire
Notes: Chapter XXIV, 116

Of that there is no manner of doubt,
No probable, possible, shadow of doubt,
No possible doubt whatever.

Gilbert, W.S.
Sullivan, Arthur
The Complete Plays of Gilbert and Sullivan
The Gondoliers
Act I

The only seasonable inquiry is, Which is of probables the most, or of improbables the least, such.

Hammond, Henry
Quoted in S. Austin Alibone's
Prose Quotations from Socrates to Macaulay
Probability

The only knowledge *a priori* is purely analytic; all empirical knowledge is probable only.

Lewis, Clarence Irving
Mind and the World-Order
Chapter X (p. 309)

There are certain notions which it is impossible to define adequately. Such notions are found to be those based on universal experience of nature. Probability is such a notion. The dictionary tells me that 'probable' means 'likely'. Further reference gives the not very helpful information that 'likely' means 'probable'.

Moroney, M.J.
Facts from Figures
The Laws of Chance (p. 4)

But is it *probable* that *probability* gives assurance?

Pascal, Blaise
The Thoughts of Blaise Pascal
Appendix: Polemical Fragments, 908

As being is to become, so is truth to belief . . . Enough if we adduce probabilities as likely as any other; for we must remember that I who am the speaker, and you who are the judges, are only mortal men, and we ought to accept the tale which is probable and enquire no further . . .

Plato
Timaeus
29

Predicted facts . . . can only be probable.

Poincaré, Henri
The Foundations of Science
Science and Hypothesis (p. 155)

I think that we shall have to get accustomed to the idea that we must
not look upon science as a 'body of knowledge' but rather as a system of
hypotheses; that is to say, as a system of guesses or anticipations which
in principle cannot be justified, but with which we work as long as they
stand up to tests, and of which we are never justified in saying that we
know that they are 'true' or 'more or less certain' or even 'probable'.

Popper, Karl R.
The Logic of Scientific Discovery (p. 317)

To say that observations of the past are certain, whereas predictions are
merely probable, is not the ultimate answer to the question of induction;
it is only a sort of intermediate answer, which is incomplete unless a
theory of probability is developed that explains what we should mean
by "probable" and on what ground we can assert probabilities.

Reichenbach, Hans
The Rise of Scientific Philosophy (p. 93)

All views are only probable, and a doctrine of probability which is not
bound to a truth dissolves into thin air. In order to describe the probable,
you must have a firm hold on the true. Therefore, before there can be
any truth whatsoever, there must be absolute truth.

Sartre, Jean-Paul
The Philosophy of Existentialism (p. 51)

It may be probable she lost it . . .

Shakespeare, William
The Complete Works of William Shakespeare
Cymbeline
Act II, Scene 4, l. 115

Most probable
That so she died . . .

Shakespeare, William
The Complete Works of William Shakespeare
Anthony and Cleopatra
Act V, Scene 2, l. 76

'Tis probable and palpable to thinking.

Shakespeare, William
The Complete Works of William Shakespeare
Othello, The Moor of Venice
Act I, Scene 2, l. 76

Which to you shall seem probable . . .

Shakespeare, William
The Complete Works of William Shakespeare
The Tempest
Act V, Scene 1, l. 249

How probable, I do not know . . .

Shakespeare, William
The Complete Works of William Shakespeare
Coriolanus
Act IV, Scene 2, l. 178

'Tis pretty, sure, and very probable . . .

Shakespeare, William
The Complete Works of William Shakespeare
As You Like It
Act III, Scene 5, l. 11

That is accounted probable which has better arguments producible for it than can be brought against it.

South, Robert
Quoted in S. Austin Alibone's
Prose Quotations from Socrates to Macaulay
Probability

The management of changes is the effort to convert certain possibles into probables, in pursuit of agreed-on preferables.

Toffler, Alvin
Future Shock (p. 407)

It is probable that many things will happen contrary to probability.

Unknown

In all the ordinary affairs of life men are used to guide their actions by this rule, namely to incline to that which is most probable and likely when they cannot attain any clear unquestionable certainty.

Wilkins, John
Of the Principles and Duties of Natural Religion (p. 30)

PROBLEMS

I have yet to see any problem, however complicated, which, when you looked at it in the right way, did not become still more complicated.

Anderson, Poul
Quoted in William Thorpe's article
Reduction v. organicism
New Scientist
Volume 43, Number 66, 25 September 1969 (p. 638)

Most problems have either many answers or no answer. Only a few problems have a single answer.

Berkeley, Edmund C.
Computers and Automation
Right Answers—A Short Guide for Obtaining Them
September 1969

Inside every large problem is a small problem struggling to get out.

Bloch, Arthur
Murphy's Law
Hoare's Law of Large Problems (p. 50)

Man is seen to be an enigma only as an individual, in mass, he is a mathematical problem.

Chambers, Robert
Vestiges of the Natural History of Creation (p. 333)

It isn't that they can't see the solution. It is that they can't see the problem.

Chesterson, Gilbert Keith
The Father Brown Omnibus
The Scandal of Father Brown
The Point of the Pin (p. 949)

. . . you're either part of the solution or part of the problem.

Cleaver, Eldridge
Speech in San Francisco, 1968

QUALITY CONTROL

MTBF *n*. [Mean Time Between Failure.] . . . Manufacturers have long been aware that too high a value for the MTBF (measured, usually, in decades or fractions of decades) leads to a stultifying sense of boredom and complacency on the part of the user.

Kelly-Bootle, Stan
The Devil's DP Dictionary

MTTR *n*. [Mean Time To Repair. Origin: *mean* "poor or inferior in grade or quality" + repair "to take off": as, Let's repair to the bar.] The possible sum of the following series, for which there is no easy convergence test:

MTTNF	Mean Time To Notice Fault
MTTRTF	Mean Time To React To Fault
MTTLFEPN	Mean Time To Locate Field Engineer's Phone Number
MTTCFE	Mean Time To Call Field Engineer
MTAFECB	Mean Time Awaiting Field Engineer's Call Back
MTTCSC	Mean Time To Check Service Contract
MTTCFES	Mean Time To Call Field Engineer's Superior
MTTLTFEDBS	Mean Time To Listen To Field Engineer's Disclaimer Blaming Software
MTTCA	Mean Time To Call Attorney
MTFFETA	Mean Time For Field Engineer To Arrive
MTTD	Mean Time To Diagnose
MTTLTFEDBS	Mean Time To Listen To Field Engineer's Disclaimer Blaming Software
MTOOSCM x M#	Mean Time Ordering/Obtaining Software/Changing Modules multiplied By Number of Modules
MTTRB	Mean Time To ReBoot
MTTRRB	Mean Time To ReReBoot

Kelly-Bootle, Stan
The Devil's DP Dictionary

The fundamental difference between engineering with and without statistics boils down to the difference between the use of a scientific method based upon the concept of laws of nature that do not allow for chance or uncertainty and a scientific method based upon the concepts of laws of probability as an attribute of nature.

Shewhart, W.A.
University of Pennsylvania Bicentennial Conference

Without quality control you, as a producer or purchaser, are in the same position as the man who bets on a horse race—with one exception, the odds are not posted.

Steadman, Frank M.
Textile World
Quality Control Posts Mill—Production Odds
Volume 94, Jul–Dec 1944 (p. 63)

You can't inspect quality into a product.

Unknown

LAST YEAR I TOOK AN 8% INCREASE IN EARNINGS, BUT GODDAMNED INFLATION WENT UP BY 11%, LEAVING ME 3% WORSE OFF

Wall Street indexes predicted nine out of the last five recessions!

Paul A. Samuelson – (See p. 156)

QUEUE

Hurry up and wait.

Old Army Saying

There is one habit which is clearly of British origin—that of queueing. Unlike the British, they have no passion for queueing; they do not like queueing for queueing's sake. But they stick to the queueing etiquette, form orderly queues at many places, and guard their rights with a morose kind of vigilance.

Mikes, George
How to be an Alien
Israel (p. 121)

RANDOMNESS

All the King's horses and all the King's men
Couldn't put Humpty Dumpty in his place again.

Carroll, Lewis
The Complete Works of Lewis Carroll
Through the Looking Glass
Humpty Dumpty

What they appear to tell us is that nothing is so alien to the human mind
as the idea of randomness.

Cohen, John
Chance, Skill, and Luck
Chapter 2, Part IV (p. 42)

The kitten was an adorable mass of silver–grey fluff and was at first
named Fluffy Ruffles through an error in sex; she was a he. But he
demonstrated such lightening changes in mood, speed, and action that
Brian remarked, "That kitten doesn't have a brain; he just has a skull
full of random numbers, and whenever he bangs his head into a chair
or ricochets off a wall, it shakes up the random numbers and causes him
to do something else."

Heinlein, Robert A.
To Sail Beyond the Sunset
Chapter 10 (p. 147)

Hardyman's Truism. Random stomping seldom catches bugs.

Peers, John
1001 Logical Laws (p. 27)

A Million Random Digits with 100,000 Normal Deviates.

The RAND Corporation
Title of Book

Iocasta. Nay, what should mortal fear, for whom the degree of fortune are supreme, and who hath clear foresight of nothing? 'Tis best to live at random, as one may.

Sophocles
The Plays of Sophocles
Oedipus the King
l. 997

Random is not haphazard

Unknown

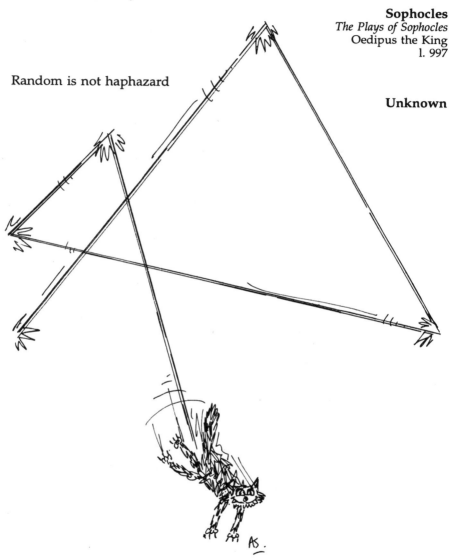

and whenever he bangs his head into a chair or ricochets off a wall, it shakes up the random numbers and causes him to do something else.

Robert A. Heinlein – (See p. 191)

REASON

Pile up facts or observations as we may, we shall be none the wiser. To learn, we must necessarily reason about what we have observed, compare the facts and judge them by other facts used as controls.

Bernard, Claude
An Introduction to the Study of Experimental Medicine (p. 16)

How easy it is for unverified assumptions to creep into our reasoning unnoticed!

Beveridge, W.I.B.
The Art of Scientific Investigation (p. 87)

Reason is the shepherd trying to corral life's vast flock of wild irrationalities.

Eldridge, Paul
Maxims for a Modern Man
2194

"Ah! my dear Watson, there we come into those realms of conjecture where the most logical mind may be at fault."

Holmes, Sherlock
in Arthur Conan Doyle's
The Complete Sherlock Holmes
The Adventure of the Empty House

"I can see nothing," said I, handing it back to my friend.

"On the contrary, Watson, you can see everything. You fail, however, to reason from what you see. You are too timid in drawing your inferences."

Holmes, Sherlock
in Arthur Conan Doyle's
The Complete Sherlock Holmes
The Adventure of the Blue Carbuncle

In solving a problem of this sort, the grand thing is to be able to reason backward. This is a very useful accomplishment, and a very easy one, but people do not practise it much . . . Most people, if you describe a train of events to them, will tell you what the result would be. They can put those events together in their minds, and argue from them that something will come to pass. There are few people, however, who, if you told them a result, would be able to evolve from their own inner consciousness what the steps were which led up to the result. This power is what I mean when I talk of reasoning backward . . .

Holmes, Sherlock
in Arthur Conan Doyle's
The Complete Sherlock Holmes
A Study in Scarlet

One of the difficulties arising out of the subjective view of probability results from the *principle of insufficient reasons*. This principle . . . holds that *if we are wholly ignorant of the different ways an event can occur and therefore have no reasonable ground for preference, it is as likely to occur one way as another.*

Kasner, Edward
Newman, James
Mathematics and Imagination (p. 229)

This kind of reasoning has weaknesses, of course, as do all forms of reasoning. If the correspondence between two things compared is, not complete, that is, if *significant* differences can be shown to exist, then the argument collapses.

Minnick, Wayne C.
The Art of Persuasion (p. 16)

My Design in this Book is not to explain the Properties of Light by Hypotheses, but to propose and prove by Reason and Experiments: In order to which I shall premise the following Definitions and Axioms.

Newton, Sir Isaac
Opticks
Book One, Part I

Reasoning goes beyond the analysis of facts.

Romanoff, Alexis L.
Encyclopedia of Thoughts
Aphorisms
1973

His reasons are as two grains of wheat hid in two bushels of chaff: you shall seek all day ere you find them, and when you have them, they are not worth the search.

Shakespeare, William
The Complete Works of William Shakespeare
The Merchant of Venice
Act I, Scene 1, l. 115

The concept of randomness arises partly from games of chance. The word 'chance' derives from the Latin *cadentia* signifying the fall of a die. The word 'random' itself comes from the French *randir* meaning to run fast or gallop.

Spencer Brown, G.
Probability and Scientific Inference
Chapter VII (p. 35)

Like all Holmes' reasoning the thing seemed simplicity itself when it was once explained.

Watson, Dr.
in Arthur Conan Doyle's
The Complete Sherlock Holmes
The Adventure of the Stockbroker's Clerk

RECURSION

recursive *adj. See* RECURSIVE.

Kelly-Bootle, Stan
The Devil's DP Dictionary

Of all ideas I have introduced to children, recursion stands out as the one idea that is particularly able to evoke an excited response.

Papert, Seymour
Mindstorm (p. 71)

To iterate is human, to recurse divine.

Unknown

REGRESSION

Where the line is to be drawn the important and the trivial cannot be settled by a formula.

Cardozo, Benjamin N.
Jacob & Youngs v. Kent, 230 *New York Reports* 239, 243, 1921

Most economists think of God as working great multiple regressions in the sky.

Fiedler, Edgar R.
Across the Board
The Three R's of Economic Forecasting—Irrational, Irrelevant and Irreverent
June 1977

Once upon a time, there was a sensible straight line who was hopelessly in love with a dot.

Juster, Norton
The Dot and the Line

How did I get into this business? Well I couldn't understand multiple regression in college, so I settled for this instead.

Caption on cartoon by unknown artist

Regression begins with the unknown and ends with the unknowable.

Unknown

Like father, like son.

Unknown

You've got to draw the line somewhere.

Unknown

The term "regression" is not a particularly happy one from the etymological point of view, but it is so firmly embedded in statistical literature that we make no attempt to replace it by an expression which would more suitably express its essential properties.

Yule, G.U.
Kendall, M.G.
An Introduction to the Theory of Statistics (p. 230)

Rowe's Rule: the odds are six to five that the light at the end of the tunnel is the headlight of an oncoming train.

Paul Dickson – (See p. 277)

RESEARCH

The way to do research is to attack the facts at the point of greatest astonishment.

Green, Celia
The Decline and Fall of Science
Aphorisms (p. 1)

Research is a way of taking calculated risks to bring about incalculable consequences.

Green, Celia
The Decline and Fall of Science
Aphorisms (p. 1)

"Research," he said, "is something that tells you that a jackass has two ears."

Lasker, Albert D.
Quoted in John Gunther's
Taken at the Flood: The Story of Albert D. Lasker (p. 96)

If you steal from one author, it's plagiarism; if you steal from many, it's research.

Mizner, Wilson
Quoted in Alva Johnson's
The Legendary Mizners
Chapter 4, The Sport (p. 66)

RESIDUALS

Almost all the greatest discoveries in astronomy have resulted from the consideration of what we have elsewhere termed RESIDUAL PHENOMENA, of a quantitative or numerical kind, that is to say, of such portions of the numerical or quantitative results of observations as remain outstanding and unaccounted for after subducting and allowing for all that would result from the strict application of known principles.

Herschel, John
Outlines of Astronomy (p. 548)

Statistics show that seventy-four per cent of wives open letters, with or without a teakettle.

Rex Stout – (See p. 258)

SAMPLE

After painstaking and careful analysis of a sample, you are always told that it is the wrong sample and doesn't apply to the problem.

Bloch, Arthur
Murphy's Law
Fourth Law of Revision (p. 48)

A person's opinion of an institution that conducts thousands of transactions every day is often determined by the one or two encounters which he has had with the institution in the course of several years.

Cochran, William G.
Sampling Techniques (p. 1)

Our knowledge, our attitudes, and our actions are based to a very large extent on samples.

Cochran, William G.
Sampling Techniques (p. 1)

Sampling is the science and art of controlling and measuring the reliability of useful statistical information through the theory of probability.

Deming, William Edwards
Some Theory of Sampling (p. 3)

A good sample-design is lost if it is not carried out according to plans.

Deming, William Edwards
Some Theory of Sampling (p. 241)

If the cost of classifying a sampling unit were zero, one could always safely recommend fantastic plans of stratified sampling, with no worry about costs. The fact is, though, that there is always a price to pay . . .

Deming, William Edwards
Sample Design in Business Research (p. 320)

I've got a little list—I've got a little list . . .
I've got *him* on the list . . .
They never would be missed—they never would be missed!

Gilbert, W.S.
Sullivan, Arthur
The Complete Plays of Gilbert and Sullivan
The Mikado
Act I

He pointed to a heap of five or six hundred letters, and laughed consumedly.

"Impossible to read them all, you know. It seemed to me that the fairest thing would be to shake them together, stick my hand in, and take out one by chance. If it didn't seem very promising, I would try a second time."

Gissing, George
New Grub Street
The Way Hither (p. 62)

One does not have to read much of the current research literature in psychology, particularly in individual and social psychology, to realize that there exists a great deal of confusion in the minds of investigators as to the necessity of obtaining a truly representative sample, describing carefully how the sample was secured, and restricting generalizations to the universe, often ill-defined, from which the sample was drawn.

McNemar, Quinn
Psychological Bulletin
Sampling in Psychological Research
Volume 37, Number 6, June 1940 (p. 33)

. . . weighting a sample appropriately is no more fudging the data than is correcting a gas volume for barometric pressure.

Mosteller, F.
Journal of the American Statistical Association
Principles of Sampling
Volume 49, Number 265, 1954 (p. 33)

Everyone who has poured a highball into the nearest potted plant after taking one sip has had some experience in sampling.

Slonim, Morris James
Sampling (p. 1)

Sampling is only one component, but undoubtedly the most important one, of that broad based field of scientific method known as statistics.

Slonim, Morris James
Sampling (p. 7)

Everybody's taken samples. When you taste a bowl of soup, you take a sample, but if you don't stir it up, it won't be a representative sample, and if you're the chef, this could yield undesirable results. It doesn't have to be a random sample, but it does have to be representative.

Unknown

The things directly observed are, almost always, only samples.

Whitehead, Alfred North
Science and the Modern World (p. 23)

OK THEORY- WE KNOW YOU'RE IN THERE - YOU'RE SURROUNDED - DROP YOUR WEAPON AND COME OUT WITH YOUR THEORETICAL HANDS UP!

Very dangerous things, theories.
Dorothy L. Sayers – (See p. 282)

SCIENCE

They who have handled the Sciences have been either Empirics or Dogmatists. The Empirics, like the Ant, amass only and use: the latter, like Spiders, spin webs out of themselves: but the course of the Bee lies midway; she gathers materials from the flowers of the garden and the field; and then by her own power turns and digests them.

Bacon, Francis
The Novum Organon
First Book, 95

The object of statistical science is to discover methods of condensing information concerning large groups of allied facts into brief and compendious expressions suitable for discussion. The possibility of doing this is based on the constancy and continuity with which objects of the same species are found to vary.

Galton, Francis
Inquiries into Human Faculty
Statistical Methods

Accordingly there are two main types of science, exact science . . . and empirical science . . . seeking laws which are generalizations from particular experiences and are verifiable (or, more strictly, 'probabilities') only by observation and experiment.

Harris, Errol E.
Hypothesis and Perception
Prevalent Views of Science (p. 25)

I am a mere street scavenger of science. With hook in hand and basket on my back, I go about the streets of science collecting whatever I find.

Magendie, François
Quoted in René Dubos'
Louis Pasteur: Free Lance of Science (p. 363)

Science does not aim, primarily, at high probabilities. It aims at a high informative content, well backed by experience. But a hypothesis may be very probable simply because it tells us nothing, or very little.

Popper, Karl R.
The Logic of Scientific Discovery (p. 399)

Facts without theory is trivia. Theory without facts is bullshit.
Unknown – (See p. 282)

STATISTICAL

A knowledge of statistical methods is not only essential for those who present statistical arguments it is also needed by those on the receiving end.

Allen, R.G.D.
Statistics for Economists
Chapter I (p. 9)

Statistical tables are essentially specific in their meaning, and they require data that are uniformly specific in the same kind and degree.

Bailey, W.B.
Cummings, John
Statistics (p. 33)

The statistical method is social mathematics par excellence.

Bell, Eric T.
The Development of Mathematics (p. 582)

Mankind in the mass is more despotically governed by the laws of chance than it ever was by the decrees of any tyrant. If our shambling race is ever to get anything but suicidal destruction out of science, it may be a necessary first step that half a dozen human beings in every hundred thousand understand the mass-reactions of creatures who, as individuals, occasionally show that they can stand erect and walk like men. To grasp and analyze mass-reactions, whether of atoms or of human beings, a mastery of the modern statistical method is essential.

Bell, Eric T.
The Development of Mathematics (p. 582)

If enough data is collected, anything may be proved by statistical methods.

Bloch, Arthur
Murphy's Law
William and Holland's Law (p. 47)

In statistical work we should be able to presume upon honesty, fidelity, and diligence.

Blodgett, James H.
Journal of the American Statistical Association
Obstacles to Accurate Statistics
New Series Number 41, March 1898 (p. 1)

We are far from having "one statistical world".

Boudreau, Frank G., MD
Kiser, Clyde V.
Problems in the Collection and Comparability of International Statistics (p. 5)

Some of the common ways of producing a false statistical argument are to quote figures without their context, omitting the cautions as to their incompleteness, or to apply them to a group of phenomena quite different to that to which they in reality relate; to take these estimates referring to only part of a group as complete; to enumerate the events favorable to an argument, omitting the other side; and to argue hastily from effect to cause, this last error being the one most often fathered on to statistics. For all these elementary mistakes in logic, statistics is held responsible.

Bowley, Arthur L.
Elements of Statistics
Part I, Chapter I (p. 13)

A statistical estimate may be good or bad, accurate or the reverse; but in almost all cases it is likely to be more accurate than a casual observer's impression, and the nature of things can only be disproved by statistical methods.

Bowley, Arthur L.
Elements of Statistics
Part I, Chapter I (p. 9)

A useful property of a test of significance is that it exerts a sobering influence on the type of experimenter who jumps to conclusions on scanty data, and who might otherwise try to make everyone excited about some sensational treatment effect that can well be ascribed to the ordinary variation in his experiment.

Cochran, William G.
Cox, Gertrude M.
Experimental Design (p. 5)

Since statistical significance is so earnestly sought and devoutly wished for by behavioral scientists, one would think that the *a priori* probability of its accomplishment would be routinely determined and well understood. Quite surprisingly, this is not the case.

Cohen, Jacob
Statistical Power Analysis for the Behavioral Sciences (p. 1)

Statistical methods of analysis are intended to aid the interpretation of data that are subject to appreciable haphazard variability.

Cox, D.R.
Hinkley, D.V.
Theoretical Statistics (p. 1)

Operational research is the application of methods of the research scientist to various rather complex practical operations . . . A paucity of numerical data with which to work is a usual characteristic of the operations to which operational research is applied.

Davies, J.T.
The Scientific Approach (p. 86)

The essence of life is statistical improbability on a colossal scale.

Dawkins, Richard
The Blind Watchmaker
Chapter 11

The method used by the scientist to find probable exact truth is what he calls "the method of least squares".

de Leeuw, A.L.
Rambling through Science
Gambling (p. 88)

You need not be a mathematical statistician to do good statistical work, but you will need the guidance of a first class mathematical statistician. A good engineer, or a good economist, or a good chemist, already has a good start, because the statistical method is only good science brought up to date by the recognition that all laws are subject to the variations which occur in nature. Your study of statistical methods will not displace any other knowledge that you have; rather, it will extend your knowledge of engineering, chemistry, or economics, and make it more useful.

Deming, William Edwards
Mechanical Engineering
Some Principles of the Shewhart Method of Quality Control
Volume 66, March 1944

The statistical method is more than an array of techniques. The statistical method is a Mode of Thought; it is Sharpened Thinking; it is Power.

Deming, William Edwards
Paper presented at meeting of the International Statistical Institute
September 1953

Statistical research is particularly necessary in the government service because of the high level of quality and economy that the public has the right to expect in government statistics.

Deming, William Edwards
Some Theory of Sampling (p. viii)

Statistical magic, like its primitive counterpart, is a mystery to the public; and like primitive magic it can never be proved wrong . . . The oracle is never wrong; a mistake merely reinforced the belief in magic.

Devons, Ely
Essays in Economics
Chapter 7 (p. 135)

There are those who are so impressed by the notion that 'quantification' is the only form of scientific knowledge, that they see no danger in the distorted, misleading, or simply ineffective picture that a statistical description of events may give. To such people the statistical picture is always to be preferred as the most meaningful and objective. It is indeed because this view is so widespread, that an argument stated in statistical terms has such a powerful influence in policy decision, and induces everyone to try to impress their case on public attention by peppering it with statistics.

Devons, Ely
Essays in Economics
Chapter 6 (p. 106)

Mr. Gradgrind sat writing in the room with the deadly statistical clock, proving something no doubt—probably, in the main, that the Good Samaritan was a bad economist.

Dickens, Charles
The Work of Charles Dickens
Hard Times
Book II, Chapter XII

I.   Thou shalt not hunt statistical significance with a shotgun.
II.   Thou shalt not enter the valley of the methods of inference without an experimental design.
III. Thou shalt not make statistical inference in the absence of a model.
IV.   Thou shalt honor the assumptions of the model.
V.   Thou shalt not adulterate thy model to obtain significant results.
VI.   Thou shalt not covet thy colleague's data.
VII. Thou shalt not bear false witness against thy control-group.
VIII. Thou shalt not worship the 0.05 significance level.
IX.   Thou shalt not apply large-sample approximations in vain.
X.   Thou shalt not infer causal relationships from statistical significance.

Driscoll, Michael F.
The American Mathematical Monthly
The Ten Commandments of Statistical Inference
Volume 84, Number 8, 1977 (p. 628)

"Try it yourself. When it asks for input, type in I, H, V, H and press the ENTER key. But you may be disappointed. There are only twenty-four possible permutations."

"Holy Seraphim. What can you do with twenty-four names of God? You think our wise men hadn't made that calculation? Read the *Sefer Jesirah*, Chapter Four, Section Sixteen. And they didn't have computers. 'Two Stones make two Houses. Three Stones make six Houses. Four Stones make twenty-four Houses. Five Stones make one hundred and twenty Houses. Six Stones make seven hundred and twenty Houses. Seven stones make five thousand and forty Houses. Beyond this point, think of what the mouth cannot say and the ear cannot hear.' You know what this is called today? Factor analysis."

Eco, Umberto
Il pendolo di Foucault (p. 35)

There comes a time in the life of a scientist when he must convince himself either that his subject is so robust from a statistical point of view that the finer points of statistical inference he adopts are irrelevant or that the precise mode of inference he adopts is satisfactory.

Edwards, A.W.F.
Likelihood (p. xi)

By applying the statistical method we cannot foretell the behavior of an individual in a crowd. We can only foretell the chance, the *probability*, that it will behave in some particular manner.

Einstein, Albert
The Evolution of Physics
Quanta (p. 299)

The primary function of a statistical consultant in a research organization is to furnish advice and guidance in the collection and use of numerical data to provide quantitative foundations for decisions.

Eisenhart, Churchill
The American Statistician
The Role of a Statistical Consultant in a Research Organization
Volume 2, Number 2, April 1948 (p. 6)

Although advice on how and when to draw graphs is available, we have no theory of statistical graphics . . .

Fienberg, Stephen E.
The American Statistician
Graphical Methods in Statistics
Volume 13, Number 4, November 1979 (p. 165)

This rather tumultuous overflow of statistical techniques from the quiet backwaters of theoretical methodology . . . into the working part of going concerns of the largest size, suggest that hidden causes have been at work . . . preparing men's minds, and shaping the institutions through which they work . . .

Fisher, Sir Ronald A.
American Scientist Magazine
The Expansion of Statistics
Volume 42, Number 2, April 1954 (p. 277)

I may be permitted to say that I never felt such a glow of loyalty and respect towards the sovereignty and magnificent sway of mathematical analysis when his answer reached me confirming, by purely mathematical reasoning, my various and laborious statistical conclusions with far more minuteness than I had dared to hope, for the original data ran somewhat roughly, and I had to smooth them out with tender caution.

Galton, Francis
Quoted in Karl Pearson's
The Life, Letters, and Labours of Francis Galton
Volume IIIA (p. 13)

Approximately half the articles published in medical journals that use statistical methods use them incorrectly.

Glantz, S.A.
Circulation
Biostatistics: How to Detect, Correct, and Prevent Errors in
the Medical Literature
Volume 61, 1980 (p. 1)

When people talk about 'the sanctity of the individual' they mean 'the sanctity of the statistical norm'.

Green, Celia
The Decline and Fall of Science
Aphorisms (p. 4)

Sometimes a David felled a Goliath of a statistical difficulty with a smooth stone. It might take a mathematician to prove how truly the stone was aimed.

Greenwood, M.
Journal of the Royal Statistical Society
Discussion (p. 522) to the paper Some Aspects of the Teaching of Statistics
Volume 102, 1939

When we can't prove our point through the use of sound reasoning, we fall back upon statistical 'mumbo jumbo' to confuse and demoralize our opponents.

Habera, Audrey
Runyon, Richard P.
General Statistics
Chapter 1 (p. 3)

Oh, the hell with!—it did not change the statistical outcome.

Heinlein, Robert A.
Time Enough for Love (p. 208)

Statistical methods are essentially methods for dealing with data that have been obtained by repetitive operations.

Hoel, P.G.
Introduction to Mathematical Statistics (p. 1)

Acceptability of a statistically *significant* result of an experiment on animal behavior in contradistinction to a result which the investigator can repeat before a critical audience naturally promotes a high output of publication. Hence, the argument that the techniques *work* has a tempting appeal to young biologists.

Hogben, Lancelot
Statistical Theory (p. 27)

And when, in pursuit of the black cat of definitive truth, more refined techniques of statistical analysis, factor analysis, and so forth, are developed, the researcher is more and more distanced from the subject of his pursuit, and the real human world in which it exists. He raises as by a sort of mathematical levitation, into that other, finer sphere, where black cats are clawless, mewless and abstract . . .

Hopkins, Harry
The Numbers Game: The Bland Totalitarianism
Chapter 7 (p. 141)

Confidence in the omnicompetence of statistical reasoning grows by what it feeds on.

Hopkins, Harry
The Numbers Game: The Bland Totalitarianism
Chapter 6 (p. 132)

Research in statistical theory and technique is necessarily mathematical, scholarly, and abstract in character, requiring some degree of leisure and detachment, and access to a good mathematical and statistical library.

Hotelling, Harold
Memorandum to the Governor of India
24 February, 1940

The purely random sample is the only kind that can be examined with entire confidence by means of statistical theory, but there is one thing wrong with it. It is so difficult and expensive to obtain for many uses that sheer cost eliminates it.

Huff, Darrell
How to Lie with Statistics (p. 21)

The use of available statistical records requires, first, that the social scientist *be familiar with the better known sources of such data* and that he *display some ingenuity in discovering less obvious material.*

Jahoda, Marie
Deutsch, Morton
Cook, Stuart
Research Methods in Social Relations
Basic Process
Part I (p. 232)

Statistical laws enable the insurance company to function, and make a profit for its shareholders. But what does statistics do for the policyholder? *Not one damn' thing!*

Jones, Raymond F.
The Non-Statistical Man (p. 32)

Sarah Bascomb was well aware that she didn't live in the same world with her husband, and that made it rather nice, she thought. It would have been exceedingly boring if they *both* talked of nothing but expectancy tables and statistical probabilities, or the PTA and young Chuck's music lessons.

Jones, Raymond F.
The Non-Statistical Man (p. 10)

. . . statistical techniques are tools of thought, and not substitutes for thought.

Kaplan, Abraham
The Conduct of Inquiry
Chapter VI, Section 29 (p. 257)

Monte Carlo method [Origin: after Count Montgomery de Carlo, Italian gambler and random number generator (1792–1838)]. A method of jazzing up the action in certain statistical and number-analytic environments by setting up a book and inviting bets on the outcome of a computation.

Kelly-Bootle, Stan
The Devil's DP Dictionary

The first mathematical discussion of the Latin Square known to modern statisticians was given by Euler in 1882. Euler does not make any specific references to previous work and merely mentions the problem as having aroused interest, but since he entitled his paper "Recherche sur une nouvelle *espèce de carré magique*" he seems to have been under the impression that the problem was fairly new . . .

Kendall, Maurice G.
The American Statistician
Who Discovered the Latin Square?
Volume 11, Number 4, August 1948 (p. 13)

Archaeologists unearthed today in Babylon a remarkable set of clay tablets recording the minutes of the 1242 annual meeting of the Babylonical Statistical Association.

King, Willford
Journal of the American Statistical Association
Consolidating Our Gains
Volume 31, Number 193, March 1936 (p. 2)

It is all too easy to notice the statistical sea that supports our thoughts and actions. If that sea loses its buoyancy, it may take a long time to regain the lost support.

Kruskal, William
The American Statistician
Coordination Today: A Disaster or a Disgrace
Volume 37, Number 3, 1983 (p. 179)

The applicability in psychology of certain of Professor R.A. Fisher's designs should be examined. Eventually, the analysis of variance will come into use in psychological research. Thus we must recognize, alongside of those natural laws which are based upon past experience without exceptions and are predicted universally, empirical generalizations admitting of possible or actual exception but nevertheless having a certain probability in the individual case. Let us call these last "statistical generalizations" since they are exhibited at their best when supported by statistical procedures.

Lewis, Clarence Irving
Mind and the World-Order
Chapter X (pp. 334–5)

. . . the statistical prediction of the future from the past cannot be generally valid, because whatever is future to any given past, is in turn past for some future. That is, whoever continually revises his judgment of the probability of a statistical generalization by its successively observed verifications and failures, cannot fail to make more successful predictions than if he should disregard the past in his anticipation of the future. This might be called the "Principle of statistical accumulation".

Lewis, Clarence Irving
Mind and the World-Order
Chapter XI (p. 386)

The statistical method is of use only to those who have found it out.

Lippmann, Walter
A Preface to Politics
The Golden Rule and After (p. 92)

No matter what the statistical problem may be, it must proceed according to a *plan*. It is always a specific question which may be answered in several more or less accurate ways. The end in view and the reasoning which can be drawn upon will indicate in which manner and within which limits the answer is to be given. According to the choice made, it may be very simple or very complicated. But under all circumstances a definite plan providing for all the detail is an absolute prerequisite.

Meitzen, August
History, Theory and Techniques of Statistics (p. 168)

No statistical judgment deals with the unit, but strictly and only with the aggregate. The variable elements of persons and things otherwise typical, that are enumerated, are always counted in a specific aggregate and under certain specific circumstances. The qualities of the objects themselves, so far as they are not typical, or the subject of the investigation, are completely unknown.

Meitzen, August
History, Theory and Techniques of Statistics (p. 163)

Statistical methods serve as land marks which point to further improvement beyond that deemed obtainable by experienced manufacturing men. Hence, after all obvious correctives have been exhausted and all normal logic indicates no further gain is to be made, statistical methods still point toward a reasonable chance for yet further gains; thereby giving the man who is doing trouble shooting sufficient courage of his convictions to cause him to continue to the ultimate gain, in spite of expressed opinion on all sides that no such gain exists.

Meyers, G.J., Jr.
Transactions, American Society of Mechanical Engineers
Discussion of E.G. Olds'
On Some of the Essentials of the Control Chart Analysis
Volume 64, July 1942

A statistical analysis, properly conducted, is a delicate dissection of uncertainties, a surgery of suppositions.

Moroney, M.J.
Facts from Figures
Statistics Undesirable (p. 3)

The organized charity, scrimped and iced,
In the name of a cautious, statistical Christ.

O'Reilly, John Boyle
In Bohemia
In Bohemia

It's been estimated that, because of the exponential growth of the world's population, between 10 and 20 percent of all the human beings who have ever lived are alive now. If this is so, does this mean that there isn't enough statistical evidence to conclusively reject the hypothesis of immortality?

Paulos, John Allen
Innumeracy (p. 99)

[Florence Nightingale] Her statistics were more than a study, they were indeed her religion. For her Quetelet was the hero as scientist, and the presentation copy of his *Physique sociale* is annotated by her on every page. Florence Nightingale believed—and in all the actions of her life acted upon the belief—that the administrator could only be successful if he were guided by statistical knowledge. The legislator—to say nothing of the politician—too often failed for want of this knowledge. Nay, she went further; she held that the universe—including human communities—were evolving in accordance with a divine plan; that it was man's business to endeavour to understand this plan and guide his actions in sympathy with it. But to understand God's thoughts, she held we must study statistics, for these are the measure of His purpose. Thus, the study of statistics was for her a religious duty.

Pearson, Karl
Life, Letters and Labours of Francis Galton
Volume II (p. 57)

There is much value in the idea of the ultimate laws being statistical laws, though why the fluctuations should be attributed to a Lucretian 'Chance', I cannot say. It is not an exactly dignified conception of the Deity to suppose him occupied solely with first moments and neglecting second and higher moments!

Pearson, Karl
The History of Statistics in the 17th and 18th Centuries against the Changing Background of Intellectual, Scientific, and Religious Thought (p. 160)

. . . it is a function of statistical method to emphasize that precise conclusions cannot be drawn from inadequate data.

Pearson, E.S.
Hartley, H.Q.
Biometrika Tables for Statisticians
Volume I (p. 83)

Statistical knowledge, though in some degree searched after in the most early ages of the world, has not till within these last 50 years become a regular object of study.

Playfair, William
Statistical Breviary

Pronouncing each word with great deliberateness, Rep. Resent asked, "Are you now, or have you ever been, a member of the American Statistical Association?"

. . .

Looking Rep. Resent straight in the eye, Minnie defiantly replied. "I refuse to answer on the grounds that it might incriminate me."

Proschan, Frank
Industrial Quality Control
Investigation of Latin Squares
Volume XI, Number 1, July 1954 (p. 31)

For the first five months they were virtually identical, but for the past four, *they showed an increasing difference!*

With shaking fingers, I worked out a Standard Deviation on the sets of totals. There was no doubt: the difference between the Centre's and CPPL's totals were significant.

Statistics don't lie . . .

Puckett, Andrew
Bloodstains (p. 80)

"Why am I surrounded," his usual understanding self today, "by statistical illiterates?"

Pynchon, Thomas
Gravity's Rainbow (p. 54)

That he must always be lovable, in need of her and never, as now, the hovering statistical cherub who's never quite been to hell but speaks as if he's one of the most fallen.

Pynchon, Thomas
Gravity's Rainbow (p. 57)

Since no scientific hypothesis is ever completely verified, in accepting a hypothesis on the basis of evidence, the scientist must make the decision that the evidence is *sufficiently* strong or that the probability is *sufficiently* high to warrant the acceptance of the hypothesis. Obviously, the decision with respect to the evidence and how strong is "strong enough" is going to be a function of the *importance*, in the typically ethical sense, of making a mistake in accepting or rejecting the hypothesis.

Rudner, R.
Scientific Monthly
Remarks on Value Judgment in Scientific Validation
Volume 79, September 1954 (p. 152)

The emergency room was a madhouse. The stormy holiday roads had yielded more than the statistical expectation of traffic accidents.

Segal, Erich
Man, Woman and Child
Chapter 26 (p. 191)

You cannot escape the statistical method, so you may as well make friends with it. You think it is cold and inhuman and impersonal, but, as a matter of fact, it is fuller of red blood and human nature than half the descriptive literature in the world.

Stamp, Josiah
Some Economic Factors in Modern Life
Chapter VIII (p. 256)

It was demonstrated however very satisfactorily, that such a ponderous mass of heterogeneous matter could not be congested and conglomerated to the nose, whilst the infant was *in Utero*, without destroying the statistical balance of the foetus, and throwing it plump upon its head nine months before the time.

Sterne, Laurence
Tristram Shandy
Book IV

But lo! men have become the tools of their tools.

Thoreau, Henry David
Walden
Economy

Factor analysis is useful especially in those domains where basic and fruitful concepts are essentially lacking and where crucial experiments have been difficult to conceive . . . They enable us to make only the crudest first map of a new domain. But if we have scientific intuition and sufficient ingenuity, the rough factorial map of a new domain will enable us to proceed beyond the factorial stage to the more direct form of psychological exploration in the laboratory.

Thurston, L.L.
Psychological Bulletin
Current Issues in Factor Analysis
Volume 37, April 1940 (p. 189)

It is not wise for a statistician who knows factor analysis to attempt problems in a science which he has not himself mastered.

Thurston, L.L.
Psychological Bulletin
Current Issues in Factor Analysis
Volume 37, April 1940 (p. 235)

A sort of question that is inevitable is: "Someone taught my students exploratory, and now (boo hoo) they want me to tell them how to assess significance or confidence for all these unusual functions of the data. (Oh, what can we do?)" To this there is an easy answer: TEACH them the JACKKNIFE.

Tukey, John W.
The American Statistician
Volume 34, Number 1, February 1980 (p. 25)

The critical ratio is Z-ness,
But when samples are small, it is *t*-ness.
Alpha means *a*,
So does *p* in a way,
And it's hard to tell *a*-ness from *p*-ness.

Unknown

The problems of statistical physics are of the greatest in our time, since they lead to a revolutionary change in our whole conception of the universe.

von Mises, Richard
Probability, Statistics, and Truth (p. 219)

I should like to give a word of warning concerning the approach to tests of significance adopted in this paper. It is very easy to devise different tests which, on the average, have similar properties, i.e., they behave satisfactorily when the null hypothesis is true and have approximately the same power of detecting departures from that hypothesis. Two such tests may, however, give very different results when applied to a given set of data. The situation leads to a good deal of contention amongst statisticians and much discredit of the science of statistics. The appalling position can easily arise in which one can get any answer one wants if only one goes around to a large enough number of statisticians.

Yates, F.
Journal of the Royal Statistical Society
Discussion on the Paper by Dr. Box and Dr. Andersen
Series B, Volume 17, 1955 (p. 31)

Statistical thinking will one day be as necessary for efficient citizenship as the ability to read and write.

Wells, H.G.
Quoted in Warren Weaver's article
Statistics
Scientific American
January 1952

She was reading birth and death statistics. Suddenly she turned to a man near her and said, "Do you know that every time I breathe a man dies?"

"Very interesting," he returned, "have you tried toothpaste?"

Jacob M. Braude –
(See p. 237)

STATISTICIAN

While, therefore, tabulation is a final process, the formulation of the scheme of tabulation should be the initial process, preceding even the formulation of the schedule, which should be determined by the character of the tables to be produced. Failure to observe this fundamental principle in statistical practice, perhaps more than any other characteristic, distinguishes the work of the amateur from that of the expert, the work of the untrained social investigator from that of the experienced scientific statistician.

Bailey, W.B.
Cummings, John
Statistics (p. 26)

He divided people into statisticians, people who knew about statistics, and people who didn't. He liked the middle group best. He didn't like the real statisticians much because they argued with him, and he thought people who didn't know any statistics were just animal life.

Balchin, Nigel
The Small Back Room (p. 137)

The statistician was let loose.

Belloc, Hilaire
The Silence of the Sea
On Statistics (p. 172)

An utterly steady, reliable woman, responsible to the point of grimness. Daisy was a statistician for the Gallup Poll.

Bellow, Saul
Herzog (p. 221)

The individual statistician must scan closely the authority on which he rests, and guard his statements with all the cautionary words which imperfect knowledge requires, or some mere child will point out the errors in his statements and his conclusions and set people wondering of what value the rest of his work may be.

Blodgett, James H.
Journal of the American Statistical Association
Obstacles to Accurate Statistics
New Series Number 41, March 1898 (p. 19)

Perhaps statisticians themselves have not always fully recognized the limitations of their work.

Bowley, Arthur L.
Elements of Statistics
Part I, Chapter I (p. 13)

Years ago a statistician might have claimed that statistics deals with the processing of data . . . to-day's statistician will be more likely to say that statistics is concerned with decision making in the face of uncertainty.

Chernoff, H.
Moses, L.E.
Elementary Decision Theory (p. 1)

[Statistics] We are constantly made aware of our awkward position; as when we are reminded that the confidence coefficient refers, after all, more to the 'lifework of the statistician' than to any particular interval. Of course sometimes we use such statements to describe how nice a job we actually have, as when we tell our students, "Look, you don't really have to be *right*, you must only be correct." Yet every day we must live with the non-zero probability that we might be that statistician who will always base his conclusions on unusual assumptions.

These thoughts have prompted the following, not at all comforting, definitions:

A statistician is a mathematician who, although he may know exactly what he is talking about and what he says may be mathematically true, may never make a correct decision.

Coole, W.P.
The American Statistician
Letters to the Editor
Volume 23, Number 1, February 1969 (p. 35)

The statistician accepts in any engagement certain responsibilities and obligations to his client and to the people that he works with. In the first place, he is the architect of a survey or experiment. It is his business to fit the various skills together to make them effective. It is important that he clarify the various responsibilities at the outset of the study.

> **Deming, William Edwards**
> *Sample Design in Business Research* (p. 10)

A statistician's responsibility is not confined to plans: he must also seek assurance of cooperation in field and office, and maintain constant touch with the work, also with the interpretation of the results.

> **Deming, William Edwards**
> *Some Theory of Sampling* (p. 8)

The minute a statistician steps into the position of the executive who must make decisions and defend them, the statistician ceases to be a statistician.

> **Deming, William Edwards**
> *Sample Design in Business Research* (p. 13)

It should be emphasized that the statistician is not necessarily abler at handling data than his colleagues trained in economics, sociology, engineering, physics, business, etc. However, because of the high transferability of the statistician's mathematical techniques, and because he acquires a broad knowledge in many fields, he is frequently adept at discovering and measuring errors in data and determining the source of the errors. He avoids drawing wrong conclusions from data whether the data be good or bad.

> **Deming, William Edwards**
> *The American Statistician*
> On the Classification of Statisticians
> Volume 2, Number 2, April 1948 (p. 16)

The only useful function of a statistician is to make predictions, and thus to provide a basis for action.

> **Deming, William Edwards**
> *Journal of the American Statistical Association*
> Quoted in W.A. Wallis'
> The Statistical Research Group, 1942–1945
> Volume 75, Number 370, June 1980 (p. 321)

[Statistician] A figure head.

> **Esar, Evan**
> *Esar's Comic Dictionary*

[Statistician] A matter-of-fact specialist.

Esar, Evan
Esar's Comic Dictionary

[Statistician] A specialist who assembles figures and then leads them astray.

Esar, Evan
Esar's Comic Dictionary

[Statistician] A man who believes figures don't lie, but admits that under analysis some of them won't stand up either.

Esar, Evan
Esar's Comic Dictionary

Too often, in many fields of science, the statistician is regarded as someone who comes on stage after data have been collected, performs standard calculations, delivers a verdict 'Significant' or 'Not Significant', and then departs.

Finney, D.J.
Statistics in Medicine
The Questioning Statistician
Volume 1, 1982 (p. 5)

The statistician cannot evade the responsibility for understanding the process he applies or recommends.

Fisher, Sir Ronald A.
The Design of Experiments (p. 1)

The statistician cannot excuse himself from the duty of getting his head clear on the principles of scientific inference, but equally no other thinking man can avoid a like obligation.

Fisher, Sir Ronald A.
The Design of Experiments (p. 2)

An Israeli statistician named Hare,
Had five factors he wished to compare,
   Levels of each were nine,
   So of course his design
Was a Hebro-Greco-Latin square.

Fleiss, Joseph L.
The American Statistician
Letters to the Editor
Volume 21, Number 4, October 1967 (p. 49)

There was a biometrician named Mabel,
Who'd never look at populations unstable.
  Using intricate relations,
  She'd find life expectations,
From the l_x's of the life table.

Fleiss, Joseph L.
The American Statistician
Letters to the Editor
Volume 21, Number 4, October 1967 (p. 49)

There was a statistician from Needham,
Who was so bright, his clients would heed him.
  Yet his embarrassed confession
  Was that, in linear regression,
He'd never subtract an extra degree of freedom.

Fleiss, Joseph L.
The American Statistician
Letters to the Editor
Volume 21, Number 4, October 1967 (p. 49)

There was a statistician from Knossus,
Who had a nonnormal neurosis.
  With techniques of newness,
  He'd measure the skewness,
And also the data's kurtosis.

Fleiss, Joseph L.
The American Statistician
Letters to the Editor
Volume 21, Number 4, October 1967 (p. 49)

We are not concerned with the very poor. They are unthinkable, and only to be appreciated by the statistician or the poet.

Forster, E.M.
Howards End
Chapter 6

Statistician—a term that is more or less equivalent to that of "Statesman".

Galton, Francis
Memories of My Life
Chapter XXI

The mathematician, the statistician, and the philosopher do different things with a theory of probability. The mathematician develops its formal consequences, the statistician applies the work of the mathematician and the philosopher describes in general terms what this application consists in. The mathematician develops symbolic tools without worrying overmuch what the tools are for; the statistician uses them; the philosopher talks about them. Each does his job better if he knows something about the work of the other two.

Good, I.J.
Science
Kinds of Probability
Volume 129, February 20, 1959 (p. 443)

Increasingly, we find ourselves caught up in the new contemporary dualism; there is the muddling-on, verbalising, impressionistic, human old world down there, and there is that Other, Finer, Rational World to which the better statisticians have already been called. Communications between the two can be tenuous.

Hopkins, Harry
The Numbers Game: The Bland Totalitarianism
Chapter 6 (p. 134)

Magruder smiled and settled back in a chair opposite Bascomb. "You are a blunt man, for a statistician," he said. "I find the uncertainties of their profession ordinarily extends to their common speech."

Jones, Raymond F.
The Non-Statistical Man (p. 29)

The early statisticians of the present century were competent at mathematics, but they were not great creative mathematicians. Karl Pearson was trained in mathematics, but Edgeworth was a classical scholar and Yule an engineer by training. Fisher, who *was* a creative mathematician, criticized his predecessors for the clumsiness of their style; but even he wrote in the tradition of English mathematics, which does not care much about extreme generalization or extreme rigor as long as it gets the right answer to its problems. The consequence was that, with few exceptions, theoretical statistics in the forties could be understood by anybody with moderate mathematical attainment, say at the first year undergraduate level. I deeply regret to say that the situation has changed so much for the worse that the journals devoted to mathematical statistics are now completely unreadable. Most statisticians deplore the fact, but there is not very much they can do about it.

Kendall, Maurice G.
Statistical Papers in Honour of George Snedecor (p. 205)

It is not primarily the responsibility of a statistician to make decisions for other people—not in general at any rate . . . It is for someone else to say what decisions should be made with [inferential] . . . information. In other words, ideally, it is the statistician's job to inform not to decide.

Kerridge, D.F.
Journal of the Royal Statistical Society
Discussion on Paper by Dr. Marshall and Professor Olkin
Series B, Volume 30, 1968 (p. 440)

An occupational hazard to which we statisticians are exposed occurs in the context of a social occasion, perhaps a dinner party. I am, let us say, seated next to a charming lady whom I have just met, and, as an initial conversational ice-breaking, she turns to me with a winning smile and says: "Now tell me what is it you do?" We must tell the truth, of course, so I reply that I am a statistician. That usually ruins a fine conversation, for in 8.6 cases out of 10 the lady's smile disappears, she turns to my rival on her other side, and I attack the fried chicken in lonely, misunderstood dignity.

Kruskal, William
American Scientist Magazine
Statistics, Molière, and Henry Adams
1967 (p. 416)

A couple of government statisticians recently threw dust on the wedding ring business by coming right out with the fact that for every male there are 1.03 females. It's about time they stop shoving the American taxpayer behind decimal points.

Miksch, W.F.
Collier's
The AVERAGE STATISTICIAN
June 17, 1950

The statistician's job is to draw general conclusions from fragmentary data. Too often the data supplied to him for analysis are not only fragmentary but positively incoherent, so that he can do next to nothing with them. Even the most kindly statistician swears heartily under his breath whenever this happens.

Moroney, M.J.
Facts from Figures
What Happens When We Take a Sample (p. 120)

There is more than a germ of truth in the suggestion that, in all society where statisticians thrive, liberty and individuality are likely to be emasculated.

Moroney, M.J.
Facts from Figures
Statistics Undesirable (p. 1)

I like to think of the constant presence in any sound Republic of two guardian angels: the Statistician and the Historian of Science. The former keeps his finger on the pulse of Humanity . . .

Sarton, George
Sarton on the History of Science
Quetelet (p. 241)

. . . as the job of finding the truth and explaining it continues to become more complex and more difficult, management again casts a doubtful eye at the statistician, for a different reason. Management's big question is no longer "What can the statistician do for us that we can't do just as well ourselves?"; the question now is, "Do our statisticians have the tools and the capacity and the experience and the persistence and the breadth of vision to seek the truth and to know it when they have found it?"

Seaton, G.L.
The American Statistician
The Statistician and Modern Management
Volume 2, Number 6, December 1948 (p. 10)

The characteristic which distinguishes the present-day professional statistician, is his interest and skill in the measurement of the fallibility of conclusions.

Snedecor, G.W.
Journal of the American Statistical Association
On a Unique Feature of Statistics
(Presidential Address to the American Statistical Association, December 1948)
Volume 44, Number 245, March 1949

I sometimes think that statisticians do not deserve quite all the hard things that are said about them. They are supposed to be cold, unemotional, bloodless and steely-eyed. But, as a matter of fact, we are all statisticians nowadays. We are either forming opinions on other people's statistics, whether we like it or not, or we are providing the raw material of statistics.

Stamp, Josiah
Some Economic Factors in Modern Life
Chapter VIII (p. 253)

Most of you would as soon be told that you are cross-eyed or knock-kneed as that you are destined to be a statistician . . .

Stamp, Josiah
Some Economic Factors in Modern Life
Chapter VIII (p. 253)

Statisticians have an understandable penchant for viewing the whole of the history of science as revolving around measurement and statistical reasoning.

Stigler, Stephen M.
The History of Statistics
Introduction (p. 1)

Everyday life is influenced more and more each day by decisions based on quantitative information. The scientific sequence—hypothesis, experiment, and test hypothesis—is now a familiar approach to problems. Only a few of all those who use it are known popularly as scientists. The distinguishing characteristic of the true scientist is not the fact that he employs scientific methodology, but rather his expertness with it.

So it is with the statistician. Nearly everyone, scientists included, draws conclusions from quantitative data. A mark of the true statistician is his special expertness at arranging an investigation and analyzing the result so as to yield the most reliable conclusions with minimum effect.

The Editors
The American Statistician
The Statistician and Everyday Affairs
Volume II, Number 5, 1948

The accounting department was working on a marketing plan for the coming year, with most of the risk evaluation work done by two employees. During a break, the subject of American history came up.

"I never realized before how close we came to losing the Revolutionary War," one commented to the other.

"What do you mean?"

"Well, they didn't understand the risks," the first one explained. "If they'd had the budget to hire a statistician, they never would have declared independence."

Thomsett, Michael C.
The Little Black Book of Business Statistics (p. 181)

Though statisticians in our time have never kept the score, Man wants a great deal here below and Women even more.

Thurber, James
Further Fables of Our Times
The Godfather and His Godchild

Too often the client (whether or not a social scientist) looks to the statistician as a man who applies the final stamp of approval—perhaps by saying, "This result is significant". Too often the statistician looks upon himself as a guardian of the proven truth . . .

Tukey, John W.
Quoted in Donald P. Ray's
Trends in Social Science
Statistical and Quantitative Methodology (p. 86)

Predictions, prophecies, and perhaps even guidance—those who suggested this title to me must have hoped for such—even though occasional indulgences in such actions by statisticians has undoubtedly contributed to the characterization of a statistician as a man who draws straight lines from insufficient data to foregone conclusions!

Tukey, John W.
Journal of the American Statistical Association
Where do We Go From Here?
Volume 55, Number 289, March 1960 (p. 80)

(The experimental statistician dare not shrink from the war cry of the analyst "Only a fool would use it, but it's better than we used to use!")

Tukey, John W.
Journal of the American Statistical Association
Unsolved Problems of Experimental Statistics
Volume 49, Number 268, December 1954 (p. 718)

The most important maxim for data analysis to heed, and one which many statisticians seem to have shunned is this: 'Far better an approximate answer to the *right* question, which is often vague, than an *exact* answer to the wrong question, which can always be made precise.' Data analysis must progress by approximate answers, at best, since its knowledge of what the problem really is will at best be approximate.

Tukey, John W.
Annals of Mathematical Statistics
The Future of Data Analysis
Volume 33, Number 1, March 1962 (pp. 13–4)

It is necessary to add that statisticians themselves are not infallible.

Unknown

If there are three statisticians on a committee, there will be 4 minority reports.

Unknown

If all the statisticians in the world were laid end to end—it would be a good thing.

Unknown

A biostatistician talks statistics to the biologist and biology to the statistician, but when he meets another biostatistician, they just discuss women.

Unknown

A statistician and the statistician's wife were marooned on a remote island. When the wife asked how they were going to escape the island and get home, the statistician replied, "Assuming we have a boat . . ."

Unknown

A statistician is a person who draws a mathematically precise line from an unwarranted assumption to a foregone conclusion.

Unknown

"Multiple births are more frequent in larger families," declares a statistician. It's mighty hard to fool these statisticians.

Unknown

Svetlana Manova, the tallest, most passionate statistician west of the Vistula, was aroused from illicit daydreams of Graeco-Latin squares by the husky voice of Bruce "Log" Linear, the new, muscular (yet intellectual) Aerobics instructor at the swank Goodness of Fit health club. It was time to start working on Log Linear's exercises, a task from which she had been deviating randomly, for reasons she could not articulate. She did not know the name of the transformation that had come over her since meeting Log, any more than she could analyze the persistent departures from normality in some of her dependants. The contrasts between Log and her were quite orthogonal, really, but factors beyond her analysis had taken possession of her life. Stretching her sleek limbs, she joined the saturated models who already lounged on the exercise floor. As the group moved through a series of unorthodox patterns on the floor (for a session with Log Linear was not restricted to hierarchical designs) she mused over the turbulent events of recent days.

Lately the relationships in her life—aside from a four-way interaction that was almost impossible to interpret—had been lacking in significance. Her last lover, the fabulously wealthy Italian recording tycoon "Disk" Riminate, had been kind and generous when they were first associated, but he had turned cruel and selfish, apparently regressing toward the mean. The other men she knew were structural zeros. Mean, square errors, she termed them. She never had any trouble rejecting their hypothesis, or eluding their predictably normal plots, but as time passed she felt she was gradually losing her residual degrees of freedom, and, as always happens in such cases, her life was becoming less significant.

And then Log had come into her life. He wasn't like the others. There was a skewness about his attitude, and at the same time a gentlemanly kurtosy, that made her want to know what it would be like to know him fully. Feeling his eyes on her as she obliquely rotated her lovely torso, she wondered how it would feel to nest her effects within his, instead of always crossing them . . .

Even then, unknown to Svetlana, Log was analyzing a related problem. He had been devoting more time and attention to statistics (her statistics) than he ever had anticipated doing. Short of doing her exercises for her, he could not have been more attentive. He knew that he needed a woman like her, but he didn't know if he—or any other man—could ever understand Manova. He feared that in the end he would be driven to a breakdown.

Unknown

It is difficult to determine what a statistician is and what a statistician is not.

Unknown

Flip a coin 100 times. Assume that 99 heads are obtained. If you ask a statistician, the response is likely to be: "It is a biased coin." But if you ask a probabilist, he may say: "Wooow, what a rare event."

Wang, Chamont
Sense and Nonsense of Statistical Inferences (p. 154)

. . . the movement of the last hundred years is all in favour of the statistician.

Wells, H.G.
The Work, Wealth and Happiness of Mankind
Chapter 9, Part 10 (p. 391)

Behind the adventurer, the speculator, comes that scavenger of adventurers, the statistician.

Wells, H.G.
The Work, Wealth and Happiness of Mankind
Chapter Nine, Part 10 (p. 390)

There is a story about two friends, who were classmates in high school, talking about their jobs. One of them became a statistician and was working on population trends. He showed a reprint to his former classmate. The reprint started, as usual with the Gaussian distribution and the statistician explained to his former classmate the meaning of the symbols for the actual population, for the average population, and so on. His classmate was a bit incredulous and was not quite sure whether to statistician was pulling his leg.

"How can you know that?" was his query. "And what is this symbol here?"

"Oh," said the statistician, "this is π."

"What is that?"

"The ratio of the circumference of the circle to its diameter."

"Well now you are pushing your joke too far," said the classmate, "surely the population has nothing to do with the circumference of the circle."

Wigner, Eugene P.
Communications in Pure and Applied Mathematics
The Unreasonable Effectiveness of Mathematics in the Natural Sciences
Volume XIII, Number 1-4, February 1960

Since the statistician can seldom or never make experiments for himself, he has to accept the data of daily experiences, and discuss as best he can the relations of a whole group of changes . . .

Yule, G.U.
Journal of the Royal Statistical Society
On the Theory of Correlation
Volume LX, December 1897 (p. 812)

CAN'T ARGUE WITH THAT!

Here Lies

HENRY SMOKETOOMUCH
COUGHED AND EXPIRED

It is now proved beyond doubt that smoking is one of the leading causes of statistics.

Fletcher Knebel –
(See p. 248)

STATISTICS

Taking for granted that the alternative to art was arithmetic, he plunged deep into statistics, fancying that education would find the surest bottom there; and the study proved the easiest he had ever approached. Even the Government volunteered unlimited statistics, endless columns of figures, bottomless averages merely for the asking. At the Statistical Bureau, Worthington Ford supplied any material that curiosity could imagine for filling the vast gaps of ignorance, and methods for applying the plasters of fact.

Adams, Henry
The Education of Henry Adams
Chapter 23 (p. 351)

No honest historian can take part with—or against—the forces he has to study. To him even the extinction of the human race should be merely a fact to be grouped with other vital statistics.

Adams, Henry
The Education of Henry Adams
Vis Interiae (p. 447)

History has never regarded itself as a science of statistics. It was the Science of Vital Energy in relation with time; and of late this radiating center of life has been steadily tending,—together with every form of physical and mechanical energy,—toward mathematical expression.

Adams, Henry
A Letter to American Teachers of History (p. 115)

. . . and you thought 'impressive' statistics were 36-24-36.

Advertisement
The American Statistician
Volume 33, Number 4, November 1979 (p. 248)

234

Statistics are the food of love.

Angell, Roger
Late Innings: A Baseball Companion
Chapter 1 (p. 9)

"Organic chemist!" said Tilley expressively. "Probably knows no statistics whatever."

Balchin, Nigel
The Small Back Room (p. 136)

[Statistics] *It is concerned with things we can count.* In so far as things, persons, are unique or ill-defined, statistics are meaningless and statisticians silenced; in so far as things are similar and definite—so many workers over 25, so many nuts and bolts made during December—they can be counted and new statistical facts are born.

Bartlett, M.S.
Essays on Probability and Statistics (p. 11)

Like dreams, statistics are a form of wish fulfillment.

Baudrillard, Jean
Cool Memories
Chapter 4

It has long been recognized by public men of all kinds . . . that statistics come under the head of lying, and that no lie is so false or inconclusive as that which is based on statistics.

Belloc, Hilaire
The Silence of the Sea
On Statistics (p. 170)

Before the curse of statistics fell upon mankind we lived a happy, innocent life, full of merriment and go, and informed by fairly good judgment.

Belloc, Hilaire
The Silence of the Sea
On Statistics (p. 171)

Statistics are the triumph of the quantitative method, and the quantitative method is the victory of sterility and death.

Belloc, Hilaire
The Silence of the Sea
On Statistics (p. 173)

As for statistics, they are given a great role in medicine, and they therefore raise a medical question which we should examine here. The first requirement in using statistics is that the facts treated shall be reduced to comparable units. Now this is very often not the case in medicine. Everyone familiar with hospitals knows what errors may mark the definitions on which statistics are based. The names of diseases are very often given haphazard, either because the diagnosis is obscure, or because the cause of death is carelessly recorded by a student who has not seen the patient, or by an employee unfamiliar with medicine. For this reason pathological statistics can be valid only when compiled from data collected by the statistician himself.

Bernard, Claude
An Introduction to the Study of Experimental Medicine (p. 136)

. . . statistics, which first secured prestige here by a supposedly impartial utterance of stark fact, have enlarged their dominion over the American consciousness by becoming the most powerful statement of the "ought"—displacers of moral imperatives, personal ideals, and unfulfilled objectives.

Boorstin, Daniel J.
The Decline of Radicalism
Chapter I

The science of statistics is the chief instrumentality through which the progress of civilization is now measured and by which its development hereafter will be largely controlled.

Boorstin, Daniel J.
The Decline of Radicalism
Chapter I

. . . statistics have tended to make facts into norms.

Boorstin, Daniel J.
The Decline of Radicalism
Chapter I

So far I speak only of impersonal statistics, which will very largely be drawn from the current facts of administration.

Booth, Charles
Charles Booth's London (p. 375)

Great numbers are not counted correctly to a unit, they are estimated; and we might perhaps point to this as a division between arithmetic and statistics, that whereas arithmetic attains exactness, statistics deals with estimates, sometimes very accurate, and very often sufficiently so for their purpose, but never mathematically exact.

Bowley, Arthur L.
Elements of Statistics
Part I, Chapter I (p. 3)

A knowledge of statistics is like a knowledge of foreign languages or of algebra; it may prove of use at any time under any circumstances.

Bowley, Arthur L.
Elements of Statistics
Part I, Chapter I (p. 4)

Statistics are for losers.

Bowman, Scotty
Sports Illustrated
A Lot More Where They Come From
April 2, 1973

She was reading birth and death statistics. Suddenly she turned to a man near her and said, "Do you know that every time I breathe a man dies?"

"Very interesting," he returned, "have you tried toothpaste?"

Braude, Jacob M.
Complete Speaker's and Toastmaster's Library
Business and Professional Pointmakers

There's too much abstract willing, purposing,
In this poor world. We talk by aggregates,
And think by systems and being used to face
Our evils in statistics, are inclined
To cap them with unreal remedies
Drawn out in haste on the other side.

Browning, Elizabeth Barrett
The Complete Poetical Works of Elizabeth Barrett Browning
Aurora Leigh
Eighth Book, l. 800

The science of *statistics*, which has only been turned to proper account in modern times, has the great honor of having proved the existence of definite rules in a number of phenomena, which had hitherto been looked upon as merely accidental or as owing their origin to an arbitrary power.

Buchner, Ludwig
Force and Matter
Free Will (p. 367)

The fundamental gospel of statistics is to push back the domain of ignorance, prejudice, rule-of-thumb, arbitrary or premature decisions, tradition, and dogmatism and to increase the domain in which decisions are made and principles are formulated on the basis of analyzed quantitative facts.

Burgess, Robert W.
Journal of the American Statistical Association
The Whole Duty of the Statistical Forecaster
Volume 32, Number 200, December 1937 (p. 636)

No matter how much reverence is paid to anything purporting to be "statistics," the term has no meaning unless the source, relevance, and truth are all checked.

Burnan, Tom
The Dictionary of Misinformation
Statistics, use, misues, and abuse of (p. 244)

. . . the worship of statistics has had the particularly unfortunate result of making the job of the plain, outright liar that much easier.

Burnan, Tom
The Dictionary of Misinformation
Statistics, use, misuse, and abuse of (p. 246)

So that I do not grossly err in facts,
Statistics, tactics, politics, and geography . . .

Byron, Lord
The Complete Poetical Works of Byron
Don Juan
Canto the Eighth, XXIV, l. 588

Statistics is a science which ought to be honourable, the basis of many most important sciences; but it is not to be carried on by steam, this science, any more than others are; a wise hand is requisite for carrying it on. Conclusive facts are inseparable from unconclusive except by a head that already understands and knows.

Carlyle, Thomas
Critical and Miscellaneous Essays
Chartism, II

Statistics, one may hope, will improve gradually, and become good for something. Meanwhile, it is to be feared the crabbed satirist was partly right, as things go: "A judicious man," says he, "looks at Statistics, not to get knowledge, but to save himself from having ignorance foisted on him."

Carlyle, Thomas
English and Other Critical Essays
Chartism, Chapter II

"And on the dead level our pace is—?" the younger suggested; for he was weak in statistics, and left all such details to his aged companion.

Carroll, Lewis
The Complete Works of Lewis Carroll
A Tangled Tale
Knot I
Elsior

. . . statistics, though not quite scripture, can be quoted by the devil.

Changing Times
Defend Yourself Against Statistics
March 1956

Beginning softly, statistics has long been handmaid to these exact sciences, apprenticed in the scullery, but now risen housekeeper, *eating with the family.*

Coats, R.H.
Journal of the American Statistical Association
Science and Society
Volume 34, Number 205, March 1939 (p. 3)

Statistics show that you have nothing to worry about.

Cogswell, Theodore R.
Quoted in Harry Harrison's
Astounding
Probability Zero (p. 329)

I have learned repeatedly, however, that the *typical* behavioral scientist approaches applied statistics with considerable uncertainty (if not actual nervousness), and requires a verbal-intuitive exposition, rich in redundancy and with many concrete illustrations.

Cohen, Jacob
Statistical Power Analysis for the Behavioral Sciences
Preface to the Original Edition

Conversation *and* statistics. Really boring.

Crichton, Michael
Rising Sun (p. 254)

Statistics are proverbially dry—forgive me if I say they are far better dry than "wet"—but to give them optimum moisture content is simply a matter of mastering fundamentals that no one should hold in contempt.

Davis, Joseph S.
Journal of the American Statistical Association
Statistics and Social Engineering
Volume 32, Number 197, March 1937 (p. 6)

Statistics are like the hieroglyphics of ancient Egypt, where the lessons of history, the precepts of wisdom, and the secrets of the future were concealed in mysterious characters.

de Jonnes, Moreau
Éléments de Statistique (p. 5)

His passion was to count everything and reduce it to statistics.

de Solla Price, Derek John
Little Science, Big Science (p. 33)

Unfortunately and inadvertently, intellectual gulfs have grown up between writers in statistics, least squares, and curve fitting. Each of the three groups has gone its own way, rediscovering developments long since discovered by the others, or—what is worse—not rediscovering them.

Deming, William Edwards
Statistical Adjustment of Data (p. iv)

I cannot oscillate a time series or properly analyse a variance . . .

Devons, Ely
Essays in Economics
Chapter 6 (p. 105)

The two most important characteristics of the language of statistics are first, that it describes things in quantitative terms, and second, that it gives this description an air of accuracy and precision.

Devons, Ely
Essays in Economics
Chapter 6 (p. 106)

The experience of falling in love could be adequately described in terms of statistics. A record of heart beats per minute, the stammering and hesitation in speech, the number of calories consumed per day, the heightening of poetic vision, measured by the number of lines of poetry written to the beloved—I won't go on; no doubt you can think of further measures.

Devons, Ely
Essays in Economics
Chapter 6 (p. 105)

How to use a language which by its very nature implies objectivity, precision and accuracy, in such a way that the subjective element of judgment, imprecision and inaccuracy are fully taken into account? It is because this task is so difficult and so rarely achieved that statistics are frequently referred to as 'the hard facts', and yet we talk of three kinds of lies—'lies, damn lies, and statistics'.

Devons, Ely
Essays in Economics
Chapter 6 (p. 111)

. . . 'statistics are only for the statistician', and even then, I might add, only for the good statistician.

Devons, Ely
Essays in Economics
Chapter 6 (p. 118)

No Chancellor of the Exchequer could introduce his proposals for monetary and fiscal policy in the House of Commons by saying 'I have looked at all the forecasts, some go one way, some another; so I decided to toss a coin and assume inflationary tendencies if it came down heads and deflationary if it came down tails' . . . And statistics, however uncertain, can apparently provide some basis.

Devons, Ely
Essays in Economics
Chapter 7 (p. 134)

What more tempting facade of rationality than the portrayal of some statistics that seem to point to policy in one direction rather than another?

Devons, Ely
Essays in Economics
Chapter 7 (p. 134)

This exaggerated influence of statistics resulting from willingness, indeed eagerness, to be impressed by the 'hard facts' provided by the 'figures', may play an important role in decision-making.

Devons, Ely
Essays in Economics
Chapter 7 (p. 134)

. . . there seems to be striking similarities between the role of economic statistics in our society and some of the functions which magic and divination play in primitive society.

Devons, Ely
Essays in Economics
Chapter 7 (p. 135)

Factual science may collect statistics, and make charts. But its predictions are, as has been well said, but past history reversed.

Dewey, John
Art as Experience
Chapter XIV (p. 346)

There are lies, damned lies, and church statistics.

Disraeli, Benjamin
Quoted in George Seldes'
The Great Quotations

In short, Statistics reigns and revels in the very heart of Physics.

Edgeworth, Francis Ysidro
Journal of the Royal Statistical Society
On the Use of the Theory of Probabilities in Statistics Relating to Society
January 1913 (p. 167)

On the other hand, the methods of statistics are so variable and uncertain, so apt to be influenced by circumstances, that it is never possible to be sure that one is operating with figures of equal weight.

Ellis, Havelock
The Dance of Life
Chapter VII, Conclusion, I (p. 273)

[Statistics] Fiction in its most uninteresting form.

Esar, Evan
Esar's Comic Dictionary

[Statistics] Data of a numerical kind looking for an argument.

Esar, Evan
Esar's Comic Dictionary

[Statistics] The science of producing unreliable facts from reliable figures.

Esar, Evan
Esar's Comic Dictionary

[Statistics] The science that can prove everything except the usefulness of statistics.

Esar, Evan
Esar's Comic Dictionary

[Statistics] The only science that enables different experts using the same figures to draw different conclusions.

Esar, Evan
Esar's Comic Dictionary

You complain that your report would be dry. The dryer the better. Statistics should be the dryest of all reading.

Farr, William
Journal of the Royal Statistical Society
Nightingale on Quetelet
Series A, 1981 (p. 144)

In the original sense of the word, 'Statistics' was the science of Statecraft: to the political arithmetician of the eighteenth century, its function was to be the eyes and ears of the central government.

Fisher, Sir Ronald A.
Sankhya
First Indian Statistical Conference, 1938
Volume 4, 1938 (p. 14)

Statistical procedure and experimental design are only two different aspects of the same whole, and that whole is the logical requirements of the complete process of adding to natural knowledge by experimentation.

Fisher, Sir Ronald A.
The Design of Experiments (p. 3)

"I was counting the waves," replied Amory gravely, "I'm going in for statistics."

Fitzgerald, F. Scott
This Side of Paradise (p. 244)

We are all victims of statistics.

Freeman, Linton C.
Elementary Applied Statistics (p. 1)

I could prove God statistically.

Gallup, George
Omni
Volume 2, Issue 2, November 1979 (p. 42)

Some people hate the very name of statistics, but I find them full of beauty and interest. Whenever they are not brutalized, but delicately handled by the higher methods, and are warily interpreted, their power of dealing with complicated phenomena is extraordinary. They are the only tools by which an opening can be cut through the formidable thicket of difficulties that bars the path of those who pursue the Science of man.

Galton, Francis
Natural Inheritance
Normal Variability (pp. 62–3)

General impressions are never to be trusted. Unfortunately when they are of long standing they become fixed rules of life, and assume a prescriptive right not to be questioned. Consequently, those who are not accustomed to original inquiry entertain a hatred and a horror of statistics. They cannot endure the idea of submitting their sacred impressions to cold-blooded verification.

Galton, Francis
Annals of Eugenics
Quote on page facing the Table of Contents

No, Mother dear, I do not *hop* into bed with every man I meet, despite your nasty little secret thoughts, but I do very much enjoy a more than occasional roll in the hay, which, if I have my statistics right, is a good deal more often than the average wife enjoys.

Gann, Ernest K.
Brain 2000 (pp. 27–8)

. . . bits of jokes, bits of statistics, bits of foolery.

Gissing, George
New Grub Street
The Sunny Way (p. 498)

Most of us have some idea of what the word *statistics* means. We should probably say that it has something to do with tables of figures, diagrams and graphs in economic and scientific publications, with the cost of living . . . and with a host of other seemingly unrelated matters of concern or unconcern . . . Our answer would be on the right lines. Nor should we be unduly upset if, to start with, we seem a little vague. Statisticians themselves disagree about the definition of the word: over a hundred definitions have been listed (W.F. Willcox, *Revue de l'Institut Internationale de Statistique*, vol. 3, p. 288, 1935).

Goodman, Richard
Modern Statistics (p. 11)

Statistics is 'hocuspocus' with numbers.

Habera, Audrey
Runyon, Richard P.
General Statistics
Chapter 1 (p. 3)

Statistics is the refuge of the uninformed.

Habera, Audrey
Runyon, Richard P.
General Statistics
Chapter 1 (p. 3)

Legal proceedings are like statistics. If you manipulate them, you can prove anything.

Hailey, Arthur
Airport
Part 3, Chapter 11 (p. 385)

Oratory is dying; a calculating age has stabbed it to the heart with innumerable dagger-thrusts of statistics.

Hancock, William Keith
Australia (p. 146)

Statistics has been likened to a telescope. The latter enables one to see further and to make clear objects which were diminished or obscured by distance. The former enables one to discern structure and relationships which were distorted by other factors or obscured by random variation.

Hand, D.J.
Psychological Medicine
The Role of Statistics in Psychiatry
Volume 15, 1985 (p. 471)

In the everyday use of statistics in business, complicated statistical methods rarely are necessary and always are to be avoided if possible. Simplicity of treatment and presentation is a requisite in the making of statistics useful in executive control.

Hayford, F. Leslie
Journal of the American Statistical Association
Some Uses of Statistics in Executive Control
Volume 31, Number 193, March 1936 (p. 36)

. . . neither statistics nor the statistician can ordinarily give the executive the final answer to his problems.

Hayford, F. Leslie
Journal of the American Statistical Association
Some Uses of Statistics in Executive Control
Volume 31, Number 193, March 1936 (p. 36)

In 1906 he started on statistics, probability, and chance by mail . . .

Heinlein, Robert A.
To Sail Beyond the Sunset
Chapter 10 (p. 147)

"Let us sit on this log at the roadside," says I, "and forget the inhumanity and ribaldry of the poets. It is in the glorious columns of ascertained facts and legalized measures that beauty is to be found. In this very log we sit upon, Mrs. Sampson," says I, "is statistics more wonderful than any poem. The rings show it was sixty years old. At the depth of two thousand feet it would become coal in three thousand years. The deepest coal mine in the world is at Killingworth, near Newcastle. A box four feet long, three feet wide, and two feet eight inches deep will hold one ton of coal. If an artery is cut, compress it above the wound. A man's leg contains thirty bones. The Tower of London was burned in 1841."

"Go on Mr. Pratt," says Mrs. Simpson. "Them ideas is so original and soothing. I think statistics are just as lovely as they can be."

Henry, O.
Tales of O. Henry
The Handbook of Hymen

"What you've got," says Idaho, "is statistics, the lowest grade of information that exists. They poison your mind . . ."

Henry, O.
Tales of O. Henry
The Handbook of Hymen

The word statistics has at least six different meanings in current use, four in the context of statistical theory alone.

Hogben, Lancelot
Science in Authority
The Present Crisis in Statistical Theory (p. 95)

For the rational study of the law the black-letter man may be the man of the present, but the man of the future is the man of statistics and the master of economics.

Holmes, O.W., Jr.
The Harvard Law Review
Path of the Law
Volume 10, 1897

Don't waste time arguing about the merits or demerits of something if you can gather some statistics that will answer the question realistically.

Hooke, Robert
Quoted in J.M. Tanur's
Statistics: A Guide to the Unknown
Statistics, Sports, and some Other Things

Do remember that your experiment is merely a hodgepodge of statistics, consisting of those cases that you happen to remember. Because these are necessarily small in number and because your memory may be biased toward one result or another, your experience may be far less dependable than a good set of statistics.

> **Hooke, Robert**
> Quoted in J.M. Tanur's
> *Statistics: A Guide to the Unknown*
> Statistics, Sports, and some Other Things

You can't argue with statistics; generally you can't even get *at* them.

> **Hopkins, Harry**
> *The Numbers Game: The Bland Totalitarianism*
> The Sterile Circle (p. 232)

As they put it in Greek, we simply don't COUNT. We consume.

> **Horace**
> *The Satires and Epistles of Horace*
> Epistle I
> To Lollius Maximus

A well-wrapped statistic is better than Hitler's "big lie"; it misleads, yet it cannot be pinned on you.

> **Huff, Darrell**
> *How to Lie with Statistics* (p. 9)

The economy was never stronger in your lifetime. But statistics must not be sedatives. Economic power is important only as it is put to human use.

> **Johnson, Lyndon B.**
> Speech at United Automobile Worker's Convention
> Atlantic City, N.J.
> 23 March, 1964

There was a time when statistics as a tool in experimentation was almost completely ignored by the experimenter; in fact, it was regarded "introducing unnecessary confusion into otherwise plain issues".

> **Johnson, Palmer O.**
> *The Scientific Monthly*
> Modern Statistical Science and its Function in
> Educational and Psychological Research
> June 1951 (p. 385)

Statistics can be used to support anything—especially statisticians.

> **Jones, Franklin P.**
> *Woman's Realm*

That was why statistics had to be invented—because people were so unstable and irrational, taken one at a time.

Jones, Raymond F.
The Non-Statistical Man (p. 15)

In statistics, you look for the common factor in order to lump otherwise dissimilar items in a single category.

Jones, Raymond F.
The Non-Statistical Man (p. 17)

The basic sequence, in ascending order, is: lies; statistics; damn statistics; benchmarks; delivery promises; DP dictionary entries.

Kelly-Bootle, Stan
The Devil's DP Dictionary

Statistics is the branch of scientific method which deals with the data obtained by counting or measuring the properties of populations of natural phenomena. In this definition 'natural phenomena' includes all the happenings of the external world, either human or not.

Kendall, Maurice G.
Stuart, A.
The Advanced Theory of Statistics
Volume I (p. 2)

It is now proved beyond doubt that smoking is one of the leading causes of statistics.

Knebel, Fletcher
Reader's Digest
December 1961

Science. I'm afraid, Dr. Noitall, you do not have any understanding of statistics.

Koshland, Daniel E., Jr.
Science
Editorial
14 January 1994

What is there about the word "statistics" that so often provokes strained silence?

Kruskal, William
American Scientist Magazine
Statistics, Molière, and Henry Adams (p. 416)

Statistics is the art of stating in precise terms that which one does not know.

Kruskal, William
American Scientist Magazine
Statistics, Molière, and Henry Adams (p. 417)

... each of us has been doing statistics all his life, in the sense that each of us has been busily reaching conclusions based on empirical observations ever since birth.

Kruskal, William
American Scientist Magazine
Statistics, Molière, and Henry Adams (p. 417)

Statistics are like alienists—they will testify for either side.

LaGuardia, Fiorello
Liberty
The Banking Investigation
May 13, 1933

He uses statistics as a drunken man uses lamp-posts—for support rather than illumination.

Lang, Andrew
Quoted in Evan Esar's
The Dictionary of Humerous Quotations

Statistics is a body of methods and theory applied to numerical evidence in making decisions in the face of uncertainty.

Lapin, Lawrence
Statistics for Modern Business Decisions (p. 2)

"I've been reading some very interesting statistics," he was saying to the other thinker.

"Ah, statistics!" said the other, "wonderful things, sir, statistics; very fond of them myself."

Leacock, Stephen
Literary Lapses
A Force of Statistics (p. 74)

. . . all statistical devices are open to abuse and require constant correction.

Lippmann, Walter
A Preface to Politics
The Golden Rule and After (p. 91)

You and I are forever at the mercy of the census-taker and the census maker. That impertinent fellow who goes from house to house is one of the real masters of the statistical situation. The other is the man who organizes the results.

Lippmann, Walter
A Preface to Politics
The Golden Rule and After (p. 92)

Statistics then is no automatic device for measuring facts.

Lippmann, Walter
A Preface to Politics
The Golden Rule and After (p. 92)

Even the most refined statistics are nothing but abstractions.

Lippmann, Walter
A Preface to Politics
The Golden Rule and After (pp. 93–4)

You cannot feed the hungry on statistics.

Lloyd George, David
Advocating Tariff Reform
Speech 1904

If for medical journals the 1960s and 1970s seem likely to be remembered as the era when the importance of ethics was emphasized, the last 20 years of this century promise to be that of statistics.

Lock, S.
Statistics in Practice

Daniel's a statistician. He sees numbers—fractions, equations, totals— and they spell out the odds for him. God knows he's brilliant at it; he's saved the lives of hundreds with those statistics.

Ludlum, Robert
The Parsifal Mosaic
Chapter 10 (p. 137)

I don't believe you. Not because you're a poor liar, but because it doesn't conform with the facts. I work with statistics, Mr. Washburn, or Mr. Bourne, or whatever your name is. I respect observable data and I can spot inaccuracies; I'm trained to do that.

Ludlum, Robert
The Bourne Identity
Chapter 9 (p. 128)

Death is a statistic for the computers.

Ludlum, Robert
The Bourne Identity
Chapter 29 (p. 401)

"There are three major and perhaps a dozen minor rental agencies, not counting the hotels, which we've covered separately. These are manageable statistics, but, of course, the garages are not."

Ludlum, Robert
The Bourne Supremacy
Chapter 18 (p. 260)

Statistics are the straw out of which I, like every other economist, have to make the bricks.

Marshall, A.
Quoted in Arthur L. Bowley's
Elements of Statistics
Part I, Chapter I (p. 8)

If you are young, then I say: Learn something about statistics as soon as you can. Don't dismiss it through ignorance or because it calls for thought . . . If you are older and already crowned with the laurels of success, see to it that those under your wing who look to you for advice are encouraged to look into this subject. In this way you will show that your arteries are not yet hardened, and you will be able to reap the benefits without doing overmuch work yourself. Whoever you are, if your work calls for the interpretation of data, you may be able to do without statistics, but you won't do as well.

Moroney, M.J.
Facts from Figures
Statistics Desirable (p. 463)

Historically, Statistics is no more than State Arithmetic, a system of computation by which differences between individuals are eliminated by the taking of an average. It has been used—indeed, still is used—to enable rulers to know just how far they may safely go in picking the pockets of their subjects.

Moroney, M.J.
Facts from Figures
Statistics Undesirable (p. 1)

Well statistics prove that you're far safer in a modern plane than in a bathtub.

Mr. Gregory
In the movie *Charlie Chan at Treasure Island*

. . . statistics refers to the methodology for the collection, presentation, and analysis of data, and for the uses of such data.

Neter, John
Wasserman, William
Applied Statistics (p. 1)

Statistics was founded by John Graunt of London, a "haberdasher of small-wares" in a tiny book called *Natural and Political Observations made upon the Bills of Mortality.*

Neuman, James R.
The World of Mathematics
Volume 3 (p. 1416)

Nobody loves a fact man. Only if the figures prove so startling a thesis that they become dramatized by their very revelation, can they be safely employed. People are skeptical of statistics. They may prove anything. The ninety-year-old patient sounded cogent enough when he assured the doctor he would never die, because statistics prove few that few men die over ninety.

Nizer, Louis
Thinking on Your Feet
Let Them In

Statistics were just as much a fantasy in their original version as in their rectified version. A great deal of the time you were expected to make them up out of your head. For example, the Ministry of Plenty's forecast had estimated the output of boots for the quarter at a hundred and forty-five millions pairs. The actual output was given as sixty-two million. Winston, however, in rewriting the forecast, marked the figure down to fifty-seven millions, so as to allow for the usual claim that the quota had been overfilled. In any case, sixty-two millions was no nearer the truth than fifty-seven millions, or a hundred and forty-five millions. Very likely no boots had been produced at all. Likelier still, nobody knew how many had been produced, much less cared.

Orwell, George
Nineteen Eighty-Four (pp. 42–3)

The fabulous statistics continued to pour out of the telescreen. As compared with last year there was more food, more clothes, more houses, more furniture, more cooking pots, more fuel, more ships, more helicopters, more books, more babies—more of everything except disease, crime, and insanity.

Orwell, George
Nineteen Eighty-Four (p. 59)

It is thus that statistics reveals more and more the inconstance and the irregularity of much social phenomena, when in lieu of applying it to a great nation altogether, one descends to a province, a town, a village.

Perrin, Jean
Quoted in Mary Jo Nye's
Molecular Reality: A Perspective on the Scientific Work of Jean Perrin (p. 25)

No study is less alluring or more dry and tedious than statistics, unless the mind and imagination are set to or that the person studying is particularly interested in the subject; which last can seldom be the case with young men in any rank of life.

Playfair, William
The Statistical Breviary (p. 16)

"Statistics" as a plural means to us simply numbers, or more particularly, number of things, and there is no acceptable synonym.

Porter, Theodore M.
The Rise of Statistical Thinking 1820–1900 (p. 11)

Statistics derives from a German term, *Statistik*, first used as a substantive by the Göttingen professor Gottfried Achenwall in 1749.

Porter, Theodore M.
The Rise of Statistical Thinking 1820–1900 (p. 23)

As the statists thinks, the bell clinks!

Proverb

La estadistics, otra mas que nos engaña.
[Statistics, yet another mistress to deceive us.]

Proverb, Spanish

They were in monthly columns. I added them and then compared the two tables.

Well, there was a difference, and a difference on the right side, more blood packs had been separated in the Centre than plasma packs had arrived in CPPL, but it wasn't as large as I would have thought. I stared at the figures for a moment, then I worked out a statistical error rate on them. The difference between them was not significant; it could be explained by random error.

Statistics don't lie, not in the right hands.

Puckett, Andrew
Bloodstains (p. 79)

"You got a ninety percent chance," he said.

Osno said quickly, "How do you get that figure?" He always did that whenever somebody pulled a statistic on him. He hated statisticians.

Puzo, Mario
Fools Die: A Novel
Chapter 24 (p. 270)

"I'm sorry. That's the Monte Carlo Fallacy. No matter how many have fallen inside a particular square, the odds remain the same as they always were. Each hit is independent of all the others. Bombs are not dogs. No link. No memory. No conditioning."

Pynchon, Thomas
Gravity's Rainbow (p. 56)

The political practice of citing only agreeable statistics can never settle economic arguments.

Ramsey, James B.
Economic Forecasting—Models or Market? (p. 77)

. . . statistics—whatever their mathematical sophistication and elegance—cannot make bad variables into good ones.

Reynolds, H.T.
Analysis of Nominal Data (p. 8)

In this country the statistical side of criminology is very imperfectly developed, and while the same cannot be said with equal force of other English-speaking countries, it yet remains true that the statistical terminology of this social science is characterized, so far as the English language is concerned, by great vagueness and uncertainty.

Robinson, Lewis Newton
History and Organization of Criminal Statistics in the United States (p. 1)

The government keeps statistics on every known thing. But there is yet to be a statistics on how many laws we are living under.

Rogers, Will
The Writings of Will Rogers
Volume IV-1 (p. 167)

Everything is figured out down to a Gnat's tooth according to some kind of statistics.

Rogers, Will
The Writings of Will Rogers
Volume IV-3 (p. 254)

Statistics, ideally, are accurate laws about large groups; they differ from other laws only in being about groups, not about individuals.

Russell, Bertrand A.
The Analysis of Matter
Chapter XIX (p. 191)

After 17 years of interacting with physicians, I have come to realize that many of them are adherents of a religion they call *Statistics*. Statistics refers to the seeking out and interpretation of p values. Like any good religion, it involves vague mysteries capable of contradictory and irrational interpretation. It has a priesthood and a class of mendicant friars. And it provides Salvation: Proper invocation of the religious dogmas of Statistics will result in publication in prestigious journals.

Salsburg, David S.
The American Statistician
The Religion of Statistics as Practiced in Medical Journals
Volume 39, Number 3, August 1985 (p. 220)

[Statistics] The art of dealing with vagueness and with interpersonal difference in decision situations.

Savage, L.J.
The Foundation of Statistics (p. 154)

History is statistics in a state of progression; statistics is history at a stand.

Schlozer, Ludwig
Westminster Review
Art. II
Transactions of the Statistical Society of London
Footnote on page 72
Volume I, Part I
April–August 1838

History is for him continuous statistics, statistics stationary history.

Schlozer, Ludwig
Quoted in August Meitzen's
History, Theory, and Techniques of Statistics (p. 37)

"How are you, Mrs. Coleman?"

"Not too bad. How's yer statistics?"

Segal, Erich
Man, Woman and Child
Chapter 1, (p. 8)

He turned over on his side and picked up the *American Journal of Statistics.* Better than a sleeping pill. He idly leafed through a particularly unoriginal piece on stochastic processes, and thought, Christ, I've said this stuff a million times. And then he realized that he himself was the author.

Segal, Erich
Man, Woman and Child
Chapter 5 (p. 42)

"I mean, here you are a professor of statistics."

"So?"

"So you have one lousy affair in your whole life. For a few lousy days. And you get a kid as evidence. Christ, what are the odds of that happening to anybody?"

"Oh," said Bob bitterly, "about a billion to one."

Segal, Erich
Man, Woman and Child
Chapter 13 (p. 109)

"My husband's a professor at M.I.T."

"Really? What's his field?"

"Statistics."

"Oh, a real brain. I'm always self-conscious when I meet that sort of mind. I can barely add a column of figures."

"Neither can Bob." Shelia smiled. "That's my job every month."

Segal, Erich
Man, Woman and Child
Chapter 17 (p. 132)

"I am Professor Beckworth," he pronounced in a kind of soprano–baritone. "Would you like to ask me some statistics, sir?"

"Yes," replied Bob. "What are the chances of this damn rain stopping today, Professor?"

"Mmm," said Jean-Claude, pondering earnestly, "You'll have to see me tomorrow about that."

Segal, Erich
Man, Woman and Child (p. 178)

We ask for no statistics of the killed,
For nothing political impinges on
This single Casualty, or all those gone,
Missing or healing, sinking or dispersed,
Hundreds of thousands counted, millions lost.

Shapiro, Karl
Collected Poems 1940–1978
Elegy for a Dead Soldier
l. 49–53

For I am one of the unpraised, unrewarded millions without whom Statistics would be a bankrupt science. It is we who are born, who marry, who die, in constant ratios.

Smith, Logan Pearsall
Trivia
Book II
Where Do I Come In?

Lawyers like words and dislike statistics.

Smith, Reginald H.
American Bar Association Journal
A Sequel: The Bar is Not Overcrowded
Volume 45, September 1959 (p. 945)

Statistics, like chloroform, lull many people to sleep in blissful ignorance. Commenting upon this human frailty to rely too much upon the logic of statistics, Dr. Jay B. Nash of New York University gives us the following story.

An inebriate lay at night in a hotel which had a sprinkler system in the room as a fire safety device, and under the glass on the dresser were the statistics on how many people had slept with peace in the room, the hours of sleep and all the other details. After reading this several times he sauntered off to bed saying,

> Now, I lay me down to sleep, statistics make my slumber sweet
> If I die, I am not concerned,
> I may get wet, but I won't get burned.

Look behind statistics! Find out how they're made up and on what definitions they are based. Don't take them at face value.

Solomon, Ben
Quoted in M. Dale Baughman's
Teacher's Treasury of Stories for Every Occasion
Youth Leader Digest

A single death is a tragedy, a million deaths is a statistic.

Stalin, Josef
cited by Anne Fremantle in
The New York Times Book Review
Unwritten Pages at the End of the Diary (p. 3)
September 28, 1958

I propose that infinitely refutable statistics be declared the official language of politics.

Stamaty, Mark Alan
Time
Washingtoon (p. 21)
September 25, 1995

Statistics is the art of lying by means of figures.

Stekel, Wilhelm
Marriage at the Crossroads
Chapter II (p. 20)

. . . elementary statistics texts tell us that the method of least squares was first discovered about 1805. Whether it had one or two or more discoverers can be argued; still the method dates from no later than 1805. We also read that Sir Francis Galton discovered regression about 1885, in studies of heredity. Already we have a puzzle—a modern course in regression analysis is concerned almost entirely with the method of least squares and its variations. How could the core of such a course date from both 1805 and 1885? Is there more than one way a sum of squared deviations can be made small?

Stigler, Stephen M.
The History of Statistics
Introduction (p. 2)

There are two kinds of statistics, the kind you look up and the kind you make up.

Stout, Rex
Death of a Doxy (p. 90)

Statistics show that seventy-four per cent of wives open letters, with or without a teakettle.

Stout, Rex
Death of a Doxy (p. 120)

Statistics are the heart of democracy.

Strunsky, Simeon
Topics of the Times
November 30, 1944

Everything is quiet, peaceful and against it all is only the silent protest of statistics . . .

Tchekhov, Anton
Tchekhov's Plays and Stories
Gooseberries

The statistics mongers . . . have calculated to a nicety how many quarter loaves, bars of iron, pigs of lead, sacks of wool, Turks, Quakers, Methodists, Jews, Catholics, and Church-of-England men are consumed or produced in the different countries of this wicked world.

Thackery, William M.
Character Sketches
Captain Rook and Mr. Pigeon

To some people, statistics is "quartered pies, cute little battleships and tapering rows of sturdy soldiers in diversified uniforms". To others, it is columns and columns of numerical facts. Many regard it as a branch of economics. The beginning student of the subject considers it to be largely mathematics.

The Editors
The American Statistician
Statistics, The Physical Sciences and Engineering
Volume 2, Number 4, August 1948

The president always led off meetings with a dizzying array of projections. Future sales would skyrocket, profits would grow, and the company would soon be a national success story. A new manager, impressed with the apparent growth potential for the company, asked one veteran executive how accurate the president's statistics were. The executive replied, "Drop a few zeros off the sales figures and put a negative sign in front of the profit projections—and you'll get a pretty good idea of where we're going."

Thomsett, Michael C.
The Little Black Book of Business Statistics (p. 6)

"This used to be a profitable company," the president complained. "But we've lost money for the last three years. What do I tell the stockholders?"

"Well," one executive piped up, "it's true that our three-year average is poor. But why cite performance? Let's blame it on statistics."

Thomsett, Michael C.
The Little Black Book of Business Statistics (p. 21)

While he is racing to the hole, the shortstop is figuring: Based on the speed of the runners and how hard the ball is hit, he probably has no chance of a double play; he may have a little chance of a play at second; and he almost certainly has no play at first. He throws to third because the distance from the hole to the bag is short, his calculation of the various probabilities led him to conclude that this was his "percentage play".

Now not so much as a glimmer of any number entered the shortstop's head in this time, yet he *was* thinking statistically.

Thorn, John
The Hidden Game of Baseball (p. 5)

I'm a woman. I'm a black woman. I'm a poor woman. I'm a fat woman. I'm a middle-aged woman. And I'm on welfare. In this country, if you're any one of those things you count less as a person. If you're all those things, you just don't count, except as a statistic.

Tillmon, Johnnie
Quoted in Francine Klagsbrun's
The First Ms. Reader
Welfare Is a Woman's Issue (p. 51)

I am a statistic.

Tillmon, Johnnie
Quoted in Francine Klagsbrun's
The First Ms. Reader
Welfare Is a Woman's Issue (p. 51)

We have no statistics to tell us whether there be any such disproportion in class where men do not die early from overwork.

Trollope, Anthony
The Eustace Diamond
XXIV

As one of the legislators of the country I am prepared to state that statistics are always false.

Trollope, Anthony
The Eustace Diamond
XXIV

Statistics is the science, the art, the philosophy, and the technique of making inferences from the particular to the general.

Tukey, John W.
Research Operations in Industry

I was deducing from the above that I had been slowing down steadily in these thirty-six years, but I perceive that my statistics have a defect: 3,000 words in the spring of 1868, when I was working seven or eight or nine hours at a sitting, has little or no advantage over the sitting of today, covering half the time and producing half the output. Figures often beguile me, particularly when I have the arranging of them myself; in which case the remark attributed to Disraeli would often apply with justice and force: "There are three kinds of lies: lies, damned lies, and statistics."

Twain, Mark
The Autobiography of Mark Twain
Chapter 29

July 4. Statistics show that we lose more fools on this day than in all the other days of the year put together.

Twain, Mark
Pudd'nhead Wilson
Chapter XVII (p. 164)

Personally, I never care for fiction or story-books. What I like to read about are facts and statistics of any kind.

Twain, Mark
Quoted in Rudyard Kipling's
From Sea to Sea
An Interview with Mark Twain

Statistics can provide a ready proof
For doubtful facts which ought to stay aloof.

Unknown
Quoted in Alexis L. Romanoff's
Encyclopedia of Thoughts

If the statistics show a trend or change, they are probably wrong.

Unknown

This seems to be one of the many cases in which the admitted accuracy of statistical processes is allowed to throw a wholly inadmissible appearance of authority over the results obtained from them. Statistics may be compared to a mill of exquisite workmanship, which grinds you stuff of any degree of fineness; but, nevertheless, what you get out depends on what you put in, and as the grandest mill in the world will not extract wheat flour from peascods, so pages of formulas will not get a definite result out of loose data.

Unknown
Paraphrase of Thomas Henry Huxley in
Quarterly Journal of Geological Society
Volume 25 1869

Statistics is the science which uses easy words for hard ideas.

Unknown

Nos numerus sumus et fruges consumere nati.
[We are just statistics, born to consume resources.]

Unknown

If I had only one day left to live, I would live it in my statistics class—it would seem so much longer.

Unknown

Statistics must be based upon something, but I'm not certain what it is.

Unknown

The beginning of modern statistics is also the beginning of modern calamity.

Unknown

What statisticians have in their briefcases is terrifying.

Unknown

The Durbin–Whatzit statistics is used to test unknown assumptions.

Unknown

Statistics prove
Near and Far
That folks who
Drive like crazy
—Are!
    Burma Shave

Unknown

Medical statistics are like a bikini. What they reveal is interesting but what they conceal is vital.

Unknown
Quoted in
The Macmillan Dictionary of Quotations

Thinking has its place . . . but, only when one is confronted with known facts and statistics. When you're in the unknown and the dark . . . you surrender your thinking in trust to the feelings that come to you out of the bush.

Van der Post, Laurens
A Far-off Place (p. 248)

There are three kinds of lies: white lies, which are justifiable; common lies—these have no justification; and statistics.

> **von Mises, Richard**
> *Probability, Statistics and Truth*
> First Lecture (p. 1)

Statistics justify and scholars seize
The salients of colonial policy.

> **Walcott, Derek**
> *Collected Poems*
> A Far Cry from Africa
> l. 7–8

Statistics provides a quantitative example of the scientific process usually described qualitatively by saying that scientists observe nature, study the measurements, postulate models to predict new measurements, and validate the model by the success of prediction.

> **Walker, Marshall**
> *The Nature of Scientific Thought* (p. 46)

Mathematical statistics provides an exceptionally clear example of the relationship between mathematics and the external world. The external world provides the experimentally measured distribution curve; mathematics provides the equation (the mathematical model) that corresponds to the empirical curve. The statistician may be guided by a thought experiment in finding the corresponding equation.

> **Walker, Marshall**
> *The Nature of Scientific Thought* (p. 50)

As a matter of fact, the whole notion of "statistical inference" often is more of a plague and less of a blessing to research workers.

> **Wang, Chamont**
> *Sense and Nonsense of Statistical Inference* (p. 29)

Statistics as a science is to quantify *uncertainty*, not *unknown*.

> **Wang, Chamont**
> *Sense and Nonsense of Statistical Inference* (p. 29)

O god thou has appointed three score years and ten as man's allotted span but O god statistics go to prove that comparatively few ever attain that age . . .

> **Waugh, Evelyn**
> In Mark Amory's
> *The Letters of Evelyn Waugh*
> Letter to Laura Herbert, dated October 1935 (p. 99)

The pretensions advanced for statistics by the student of it are undoubtedly gaining increased authority with the public.

Westminster Review
Art II
Transactions of the Statistical Society of London, Volume I, Part I
Volume 29, 1838 (p. 45)

Statistics has been called a science. It is said to connect its facts by a chain of causation: if it did so, it would be a science, though even then not a distinct and separate science. But the observations of astronomy may be called the science of astronomy as properly as statistics may be denominated a science. No mere record and arrangement of facts can constitute a science.

Westminster Review
Art II
Transactions of the Statistical Society of London, Volume I, Part I
Volume 29, 1838 (p. 69)

But statistics is not a science, and cannot be one. Studied as the statistical council have decreed statistics shall be studied, no department of human knowledge ever could become a science—a collection of theories— because they have put their veto on theorizing. But statistics is not even a department of human knowledge; it is merely a form of knowledge—a mode of arranging and stating facts which belong to various sciences.

Westminster Review
Art II
Transactions of the Statistical Society of London, Volume I, Part I
Volume 29, 1838 (p. 70)

Just as data gathered by an incompetent observer are worthless—or by a biased observer, unless the bias can be measured and eliminated from the result—so also conclusions obtained from even the best data by one unacquainted with the principles of statistics must be of doubtful value.

White, William F.
A Scrap-Book of Elementary Mathematics
The Mathematical Treatment of Statistics (p. 156)

There is a curious misconception that somehow the mathematical mysteries of Statistics help Positivism to evade its proper limitation to the observed past. But statistics tell you nothing about the future unless you make the assumption of the permanence of statistical form. For example, in order to use statistics for prediction, assumptions are wanted as to the stability of the mean, the mode, the probable error, and the symmetry or skewness of the statistical expression of functional correlation.

Whitehead, Alfred North
Adventures of Ideas
Cosmologies
Section IV

Figures may not lie, but statistics compiled unscientifically and analyzed incompetently are almost sure to be misleading, and when this condition is unnecessarily chronic the so-called statisticians may be called liars.

Wilson, E.B.
Bulletin of the American Mathematical Society
Volume 18, 1912

But in all cases remember that *statistics is not a spectator sport.*

Wonnacott, Ronald J.
Wonnacott, Thomas H.
Introductory Statistics (p. 5)

"Those Platonists are a curse," he said,
"God's fire upon the wan,
A diagram hung there instead,
More women born than men."

Yeats, William Butler
The Collected Poems of W.B. Yeats
Statistics

SURVEYS AND
QUESTIONNAIRES

A questionnaire is never perfect: some are simply better than others.

Deming, William Edwards
Some Theory of Sampling (p. 31)

A perfect survey is a myth.

Deming, William Edwards
Some Theory of Sampling (p. 24)

The only excuse for taking a survey is to enable a rational decision to be made on some problem that has arisen and on which decision, right or wrong, will be made.

Deming, William Edwards
Some Theory of Sampling (p. 545)

... neither the interviewer nor the instrument should act in any way upon the situation. The question, ideally, should be so put and so worded as to be unaffected by contextual contaminations. The interviewer must be an inert agent who exerts no influence or response by tone, expression, stance, or statement. The question must be unloaded in that it does not hint in any way that one response is more desirable or more correct than any other response. It must be placed in the sequence of the instrument in such a way that the subject's response is not affected by previous queries or by his own previous responses.

Deutscher, I.
Quoted in S.Z. Nagi and R.G. Corwin's
The Social Contexts of Research
Public and Private Opinions: Social Situations and Multiple Realities

No aphorism is more frequently repeated with field trial, than that we must ask Nature few questions or, ideally, one question at a time.

The writer is convinced that this view is wholly mistaken. Nature, he suggests, will best respond to a logical and carefully thought out questionnaire; indeed, if we ask her a single question, she will often refuse to answer until some other topic has been discussed.

Fisher, Sir Ronald A.
Journal of the Ministry of Agriculture of Great Britain
Volume 33 (p. 511)

"But what is the purpose of your survey?" he asked.

"Does it need a purpose? I tell you, I just made it up."

"But your numbers are too few to be significant. You can't fair a curve with so little data. Besides, your conditions are uncontrolled. your results don't mean anything."

Heinlein, Robert A.
Beyond This Horizon (p. 6)

The time of busy people is sometimes wasted by time-consuming questionnaires dealing with inconsequential topics, worded so as to lead to worthless replies, and circulated by untrained and inexperienced individuals, lacking in facilities for summarizing and disseminating any worthwhile information which they may obtain.

Norton, John K.
Quoted in Douglas R. Berdie and John F. Anderson's
Questionnaires: Design and Use (p. ix)

A questionnaire is not just a list of questions or a form to be filled out. It is essentially a scientific instrument for measurement and for collection of particular kinds of data. Like all such instruments, it has to be specifically designed according to particular specifications and with specific aims in mind, and the data it yields are subject to error. We cannot judge a questionnaire as good or bad, efficient or inefficient, unless we know what job it was meant to do. This means that we have to think not merely about the wording of particular questions, but first and foremost, about the design of the investigation as a whole.

Oppenheim, Abraham Naffali
Questionnaire Design and Attitude Measurement (pp. 2–3)

Your sales last year just paralleled the sales of rum cokes in Rio de Janeiro, as modified by the sum of the last digits of all new telephone numbers in Toronto. So, why bother with surveys of your own market? Just send away for the data from Canada and Brazil.

Strong, Lydia
Management Review
Sales Forecasting: Problems and Prospects
September 1956 (p. 803)

SYMMETRY

Equiprobability in the physical world is purely a hypothesis. We may exercise the greatest care and the most accurate of scientific instruments to determine whether or not a penny is symmetrical. Even if we are satisfied that it is, and that our evidence on that point is conclusive, our knowledge, or rather our ignorance, about the vast number of other causes which affect the fall of the penny is so abysmal that the fact of the penny's symmetry is a mere detail. Thus, the statement "head and tail are equiprobable" is at best an assumption.

Kasner, Edward
Newman, James
Mathematics and the Imagination (p. 251)

Theory like mist on eyeglasses. Obscure facts.

Charlie Chan –
(See p. 276)

TABLES

Tables are like cobwebs, like the sieve of the Danaides; beautifully reticulated, orderly to look upon, but which will hold no conclusion. Tables are abstractions . . . There are innumerable circumstances; and one circumstance left out may be the vital one on which all turned . . . Conclusive facts are inseparable from inconclusive except by a head that already understands and knows.

Carlyle, Thomas
English and other Critical Essays
Chartism
Chapter II

The way statistics are presented, their arrangement in a particular way in tables, the juxtaposition of sets of figures, in itself reflects the judgment of the author about what is significant and what is trivial in the situation which the statistics portray.

Devons, Ely
Essays in Economics
Chapter 6 (p. 109)

. . . some witty comments made by A.L. Bowley . . .
Footnote: (a) The terms used in the headings and margins of the table are all employed in a technical sense, known only to the officers who compiled it, and which they are unable for official reasons to divulge. (b) The sub-divisions of the table and the region to which it refers have been changed since the last return was published. (c) Before tabulation the data have been subjected to numerous adjustments, allowances and other corrections, of a kind to vitiate any tests of significance which the reader may be tempted to apply to them.

Fisher, Sir Ronald A.
Sankhya
Presidential Address
First Indian Statistical Conference, 1938
Volume 4, 1938 (p. 15)

Worst of all, however, are journals that publish tables giving the results, mostly unintelligible, of multiple range tests, the said results receiving no mention in the text. The last fault arises possibly from the misconceived idea that the property of significance resides in the data themselves, not in the contrasts they estimate. Accordingly, if the data are "significant," the author is free to comment on any feature however trivial; if they are not, interpretation is deemed impermissible.

Peirce, Charles Sanders
Quoted in Samuel Kotz and Norman L. Johnson's
Breakthroughs in Statistics
Volume II (p. 61)

Information that is imperfectly acquired, is generally as imperfectly retained; and a man who has carefully investigated a printed table, finds, when done, that he has only a very faint and partial idea of what he has read; and that like a figure imprinted on sand, is soon totally erased and defaced.

Playfair, William
The Commercial and Political Atlas (p. 3)

Education is an admirable thing, but it is well to remember from time to time that nothing that is worth knowing can be taught.

Oscar Wilde – (See p. 272)

TEACHING

When an engineer apologetically approaches a statistician, graph in hand, and asks how he should fit a straight line to these points, the situation is not unlike the moment when one's daughter inquires where babies come from. There is a need for tact, there is a need for delicacy, but here is opportunity for enlightenment and it must not be discarded casually—or destroyed with the glib answer.

Acton, F.S.
National Bureau of Standards Report 12-10-51 (p. 1)

To teach doubt and Experiment
Certainly was not what Christ meant.

Blake, William
The Complete Writings of William Blake
The Everlasting Gospel
d, l. 49

The teaching of probabilistic reasoning, so very common and important a feature of modern science, is hardly developed in our educational system before college.

Bruner, Jerome Seymour
The Process of Education (p. 45)

Statistics is not the easiest subject to teach, and there are those to whom anything savoring of mathematics is regarded as for ever anathema.

Moroney, M.J.
Facts from Figures
Statistics Desirable (p. 458)

271

It is hard to understand why he failed to appreciate the pedagogical value of designing an experiment to illustrate a point of theory, predicting the result, running the experiment, and then taking the consequences if it turned out wrong.

Olds, Edwin G.
Journal of the American Statistical Association
Teaching Statistical Quality Control for Town and Gown
Volume 44, 1949 (pp. 223–4)

Two managers were taking a course in basic statistics. After an evening in class, one said to the other, "I've noticed that every time a new idea is introduced, I have to look up three or four words just to make sense out of the idea. Why do statisticians obfuscate their message with so much terminology? Why can't they simplify it instead?"

The second manager replied, "I'll tell you, but only after I find out what 'obfuscate' means."

Thomsett, Michael C.
The Little Black Book of Business Statistics (p. 117)

Teaching data analysis is not easy, and the time allowed is always far from sufficient.

Tukey, John W.
Annals of Mathematical Statistics
The Future of Data Analysis
Volume 33, Number 1, March 1962 (p. 11)

Education is an admirable thing, but it is well to remember from time to time that nothing that is worth knowing can be taught.

Wilde, Oscar
Epigrams: Phrases and Philosophies for the Use of the Young
Sebastian Melmoth

TESTING

Rejection rules are not significance tests.

Anscombe, F.J.
Technometrics
Rejection of Outliers
Volume 2, 1960 (p. 126)

There is no more pressing need in connection with the examination of experimental results than to test whether a given body of data is or is not in agreement with any suggested hypothesis.

Fisher, Sir Ronald A.
Statistical Methods for Research Workers (p. 250)

Comparisons do ofttime great grievance.

Lydgate, John
Bochas
Book III, Chapter VIII

No instrument smaller than the World is fit to measure men and women: Examinations measure Examinees.

Raleigh, Sir Walter
Laughter from a Cloud
Some Thoughts on Examinations (p. 120)

In an examination those who do not wish to know ask questions of those who cannot tell.

Raleigh, Sir Walter
Laughter from a Cloud
Some Thoughts on Examinations (p. 120)

Beware of the confounded effect!

Unknown

Examinations are pure humbug from beginning to end.

Wilde, Oscar
Epigrams: Phrases and Philosophies for the Use of the Young
Oscariana

In examinations the foolish ask questions that the wise cannot answer.

Wilde, Oscar
Epigrams: Phrases and Philosophies for the Use of the Young
Phrases and Philosophies

It is not nice to be wedded to anything—not even to a theory.
Samuel Butler – (See p. 276)

THEORY

The real beginning of the theory of probability goes back to 1654, when Pascal and Fermat laid down the fundamental principles in a short correspondence. Most mathematical disciplines have had respectable enough parents; the progenitors (not Fermat and Pascal) of the theory of probability were thoroughly disreputable.

Bell, Eric T.
Mathematics: Queen and Servant of Science (p. 377)

Shortly after they were married, one of Corde's academic friends had congratulated him, saying, "Do you remember that old piece of business from probability theory, that if a million monkeys jumped up and down on the keys of typewriters for a million years one of them would compose *Paradise Lost?*"

Bellow, Saul
The Dean's December
Chapter iv (p. 80)

The world is more complicated than most of our theories make it out to be.

Berkeley, Edmund C.
Computers and Automation
Right Answers—A Short Guide for Obtaining Them
September 1969 (p. 20)

A theory is merely a scientific idea controlled by experiment.

Bernard, Claude
An Introduction to the Study of Experimental Medicine (p. 26)

That quantity which, when multiplied by, divided by, added to, or subtracted from the answer you get, gives you the answer you should have gotten.

Bloch, Arthur
Murphy's Law
Skinner's Constant (p. 36)

Your theory is most excellent, and I shall endeavour to collect facts for you with a view to its elucidation.

Buckland, Frank
Quoted in Karl Pearson's
The Life, Letters, and Labours of Francis Galton
Volume II (p. 87)

It is not nice to be wedded to anything—not even to a theory.

Butler, Samuel
Samuel Butler's Note-Books (p. 116)

For that theory [mathematical theory of statistics] is solely concerned with working out the properties of the theoretical models, whereas what matters—and what in one sense is most difficult—is to decide what theoretical model best corresponds to the real world-situation to which statistical methods must be applied. There is a great danger that mathematical pupils will imagine that a knowledge of mathematical statistics alone makes a statistician.

Champernowne, D.G.
Journal of the Royal Statistical Society
A Discussion on the Teaching of Mathematical Statistics at the University Level
Volume 118, 1955 (p. 203)

Theory like mist on eyeglasses. Obscure facts.

Chan, Charlie
In the movie *Charlie Chan in Egypt*

But Pop, I've got a theory.

Chan, Jimmy
In the movie *Charlie Chan in Panama*

A man warmly concerned with any large theories has always a relish for applying them to any triviality.

Chesterson, Gilbert Keith
The Father Brown Omnibus
The Wisdom of Father Brown
The Absence of Mr. Glass

"I'd be glad to settle without the theory," remarked Kimball, "if I could even understand what this thing is—or what it's supposed to do."

Clarke, Arthur C.
The Lost Worlds of 2001
Chapter 30

"And that," said Kaminski, "reminds me of another quotation—one of Niels Bohr's. 'Your theory is crazy—but not crazy enough to be true.' "

Clarke, Arthur C.
The Lost Worlds of 2001
Chapter 30

Theory is worth little, unless it can explain its own phenomena, and it must effect this with out contradicting itself; therefore, the facts are sometimes assimilated to the theory, rather that the theory to the facts.

Colton, Charles Caleb
Lacon: or many things in a few words (p. 77)

Professors in every branch of the sciences prefer their own theories to truth: the reason is, that their theories are *private* property, but truth is *common stock*.

Colton, Charles Caleb
Lacon: or many things in a few words (p. 189)

. . . for without the making of theories I am convinced there would be no observation.

Darwin, Charles
The Life and Letters of Charles Darwin
Volume II
C. Darwin to C. Lyell
June 1st [1860] (p. 108)

. . . a theory arises from a leap of the imagination . . .

Davies, J.T.
The Scientific Approach (p. 11)

Theories are generalizations and unifications, and as such they cannot logically follow only from our experiences of a few particular events. Indeed we often generalize from a single event, just as a dog does who, having once seen a cat in a certain driveway, looks eagerly around whenever he passes that place in the future. Evidently this latter activity is equivalent to testing the theory . . . that "there is always a cat in that driveway".

Davies, J.T.
The Scientific Approach (p. 11)

Rowe's Rule: the odds are six to five that the light at the end of the tunnel is the headlight of an oncoming train.

Dickson, Paul
Washingtonian
November 1978

Schumper's Observation of Scientific Theories. Any theory can be made to fit the facts by means of appropriate additional assumptions.

Quoted in Paul Dickson's
The Official Rules (S-165)

We have found a strange footprint on the shores of the unknown. We have devised profound theories, one after another to account for its origin. At last, we have succeeded in reconstructing the creature that made the footprint. And lo! It is our own.

Eddington, Sir Arthur Stanley
Space, Time and Gravitation (p. 131)

A theory can be proved by experiment; but no path leads from experiment to the birth of a theory.

Einstein, Albert
The Sunday Times
18 July 1976

The possession of an original theory which has not yet been assailed must certainly sweeten the temper of a man who is not beforehand ill-natured.

Eliot, George
The Impressions of Theophrastus Such
How We Encourage Research (p. 26)

About binomial theorem I'm teeming with a lot o' news—
With many cheerful facts about the square of the hypotenuse.

Gilbert, W.S.
Sullivan, Arthur
The Complete Plays of Gilbert and Sullivan
The Pirates of Penzance
Act I

Theories that go counter to the facts of human nature are foredoomed.

Hamilton, Edith
The Roman Way
Comedy's Mirror

Facts are of not much use, considered as facts. They bewilder by their number and their apparent incoherency. Let them be digested into theory, however, and brought into mutual harmony, and it is another matter. Theory is of the essence of facts. Without theory scientific knowledge would be only worthy of the mad house.

Heaviside, Oliver
Electromagnetic Theory
Chapter I, Introduction (p. 12)

And I believe that the Binomial Theorem and a Bach Fugue are, in the long run, more important than all the battles of history.

Hilton, James
This Week Magazine
1937

One forms provisional theories and waits for time or fuller knowledge to explode them. A bad habit, Mr. Ferguson, but human nature is weak.

Holmes, Sherlock
in Arthur Conan Doyle's
The Complete Sherlock Holmes
The Adventure of the Sussex Vampire

I don't mean to deny that the evidence is in some ways very strong in favour of your theory, I only wish to point out that there are other theories possible.

Holmes, Sherlock
in Arthur Conan Doyle's
The Complete Sherlock Holmes
Adventure of the Norwood Builder

No theory is sacred.

Hubble, Edwin
The Nature of Science and other Lectures
Experiment and Experience (p. 41)

A first-rate theory predicts; a second-rate theory forbids and a third-rate theory explains after the event.

Kitaigordski, Aleksander Isaakovich
Lecture, ICU Amsterdam, August 1975

In the beginning there was de Moivre, Laplace, and many Bernoullis, and they begat limit theorems, and the wise men saw that it was good and they called it by the name of Gauss. Then there were new generations and they said that it had experimental vigor but lacked in rigor. Then came Chebyshev, Liapounov, and Markov and they begat a proof and Polya saw that it was momentous and he said that its name shall be called the Central Limit Theorem.

Then came Lindeberg and he said that it was elementary, for Taylor had expanded that which needed expansion and he said it twice, but Levy had seen that Fourier transforms are characteristic functions and he said "Let them multiply and bring forth limit theorems and stable laws." And it was good, stable, and sufficient, but they asked "Is it necessary"? Levy answered, "I shall say verily unto you that it is not necessary, but the time shall come when Gauss will have no parts except that they be in the image of Gauss himself, and then it will be necessary." It was a prophecy, and when Cramer announced that the time had come, and there was much rejoicing and Levy said it must be recorded in the bibles and he did record it, and it came to pass that there were many limit theorems and many were central and they overflowed the chronicles and this was the history of the central limit theorem.

LeCam, L.
Statistical Science
The Central Limit Theorem around 1935
Volume 1, Number 1, February 1986 (p. 86)

Three Indian women are sitting side by side. The first, sitting on a goatskin, has a son who weighs 170 pounds. The second, sitting on a deerskin, has a son who weighs 130 pounds. The third, seated on a hippopotamus hide, weighs 300 pounds. What famous theorem does this illustrate?

Naturally, the answer is that the squaw on the hippopotamus is equal to the sons of the squaws on the other two hides.

Moger, Art.
The Complete Pun Book (pp. 23–4)

The argument seemed sound enough, but when a theory collides with a fact, the result is a tragedy.

Nizer, Louis
My Life in Court
Proxy Battle (p. 433)

To refuse to consider any possibility is merely the old habit of making theory the measure of reality.

Oman, John
The Natural and the Supernatural
Chapter XV (p. 269)

The theory of probabilities is simply the science of logic quantitatively treated.

Peirce, Charles Sanders
Writings of Charles Sanders Peirce
Volume 3 (p. 278)

On another occasion I awoke covered in sweat. I had just dreamt the obvious solution to my nightmare. Standing beside a huge blackboard covered in equations, a mathematician was concluding his demonstration, in front of a turbulent audience, that the celebrated "Monte Carlo Theorem" was generalisable; that meant not just that a roulette player placing his stake on a random number had just as much chance of winning as a martingale player systematically doubling his stake on the same number on each loss in order to recoup eventually . . .

Perec, Georges
Life: A User's Manual (p. 146)

Sir, please believe me, it's the first time this has ever happened. Have another try, don't get upset. You know our Theorems are GUARANTEED.

Petit, Jean-Pierre
Euclid Rules OK (p. 11)

Coincidences, in general, are great stumbling-blocks in the way of that class of thinkers who have been educated to know nothing of the theory of probabilities—that theory to which the most glorious objects of human research are indebted for the most glorious of illustration.

Poe, Edgar Allen
Tales of Mystery and Imagination
The Murders in the Rue Morgue (p. 208)

But this is sufficient to show that a high probability cannot be one of the aims of science. For the scientist is most interested in theories with a high content. He does not care for highly probable trivialities but for bold and severely testable (and severely tested) hypotheses. If (as Carnap tell us) a high degree of confirmation is one of the things we aim at in science, then degree of confirmation cannot be identified with probability.

This may sound paradoxical to some people. But if high probability were an aim of science, then scientists should say as little as possible, and preferably utter tautologies only. But their aim is to 'advance' science, that is to *add* to its content. Yet this means lowering its probability. And in view of the high content of universal laws, it is neither surprising to find that their probability is zero . . .

Popper, Karl R.
Conjectures and Refutations: The Growth of Scientific Knowledge (p. 286)

The study of inductive inference belongs to the theory of probability, since observational facts can make a theory only probable but will never make it absolutely certain.

Reichenbach, Hans
The Rise of Scientific Philosophy (p. 231)

A theory is worthless without good supporting data.

Romanoff, Alexis L.
Encyclopedia of Thoughts
Aphorisms
2410

Very dangerous things, theories.

Sayers, Dorothy L.
The Unpleasantness at the Bellona Club
Chapter 4

It is noteworthy that the etymological root of the word *theatre* is the same as that of the word *theory*, namely a view. A theory offers us a better view.

Seeger, Raymond J.
Journal of the Washington Academy of Sciences
Volume 36, 1946 (p. 286)

If you have to prove a theorem, do not rush. First of all, understand fully what the theorem says, try to see clearly what it means. Then check the theorem; it could be false. Examine the consequences, verify as many particular instances as are needed to convince yourself of its truth. When you have satisfied yourself that the theorem is true, you start proving it.

Unknown

The Treadmill Theorem states that every solution entails k GE. 1 new problems.

Unknown

The Dirty Data Theorem states that "real world" data tends to come from bizarre and unspecifiable distributions of highly correlated variables and have unequal sample sizes, missing data points, non-independent observations, and an indeterminate number of inaccurately recorded values.

Unknown

Why bother to make it elegant if it already works.

Unknown

Facts without theory is trivia. Theory without facts is bullshit.

Unknown

The supreme misfortune is when theory outstrips performance.

Da Vinci, Leonardo
Notebooks
c

"Let us work without theorising," said Martin; "tis the only way to make life endurable."

Voltaire
Candide
XXX

Except perhaps for a few of the deepest theorems, and perhaps not even these, most of the theorems of statistics would not survive in mathematics if the subject of statistics itself were to die out. In order to survive the subject must be more responsive to the needs of application.

Wolfowitz, J.
Essays in Probability and Statistics (p. 748)

Why bother to make it elegant if it already works.
Unknown – (See p. 282)

TRUTH

Approximate truth is the only truth attainable, but at least one must strive for that, and not wade off into arbitrary falsehood.

Eliot, George
The George Eliot Letters
Volume IV (p. 43)

. . . in the statistical world you can multiply ignorance by a constant and get truth.

Jones, Raymond F.
The Non-Statistical Man (p. 58)

There are certain statements which, though they are false as hell, must be treated as though they were true gospel.

Trollope, Anthony
The Eustace Diamond
LXXVIII

VARIABILITY

McDougall's freedom was my variance. McDougall hoped that variance would always be found in specifying the laws of behavior, for there freedom might still persist. I hoped then—less wise than I think I am now (it was 31 years ago)—that science would keep pressing variance towards zero as a limit. At any rate this general fact emerges from this example: freedom, when you believe it is operating, always resides in an area of ignorance. If there is a known law, you do not have freedom.

Boring, E.G.
The Scientific Monthly
When is Human Behavior Predetermined
Volume 84, 1957 (p. 190)

It is clear that one who attempts to study precisely things that are changing must have a great deal to do with measures of change.

Cooley, Charles
Journal of the American Statistical Association
Observations on the Measure of Change
New Series, Number 21, March 1893

Variety's the very spice of life,
That gives it all its flavour.

Cowper, William
Cowper: Poetical Works
The Task
Book II (The Timepiece)
l. 606

The computer informed her that three spaces accounted for eighty-one percent of variance.

Crichton, Michael
The Terminal Man
Chapter 6 (p. 47)

Many laws regulate variation, some few of which can be dimly seen . . . I will here only allude to what may be called correlated variation. Important changes in the embryo or larva will probably entail changes in the mature animal . . . Breeders believe that long limbs are almost always accompanied by an elongated head . . . cats which are entirely white and have blue eyes are generally deaf . . . it appears that white sheep and pigs are injured by certain plants whilst dark-coloured individuals escape . . .

Darwin, Charles
The Origin of Species
Chapter I
Effects of Habit and the Use or Disuse of Parts

Before the inherent variability of the test-animals was appreciated, assays were sometimes carried out on as few as three rabbits: as one pharmacologist put it, those were the happy days.

Fieller, E.C.
Journal of the Royal Statistical Society
Volume vii (p. 3)

Two variable organs are said to be co-related when the variation of one is accompanied on the average by more or less variation of the other, and in the same direction.

Galton, Francis
Proceedings of the Royal Society of London
Co-relations and Their Measurements, Chiefly from Anthropometric Data
Volume 45, 1888

The incalculable number of petty accidents that concur to produce variability among brothers, makes it impossible to predict the exact qualities of any individual from hereditary data. But we may predict average results with great certainty . . . and we can also obtain precise information concerning the penumbra of uncertainty that attaches itself to single predictions.

Galton, Francis
Natural Inheritance
Process in Heredity (pp. 16–7)

If we knew the little differences which divide one man from another, even within the same family, we should have the key to most of life's riddles.

Galton, Francis
Quoted in Karl Pearson's
The Life, Letters, and Labours of Francis Galton
Volume I (p. 55)

. . . to me the form of the egg has never appeared to have aught to do with the engenderment of the chick, but to be a mere accident; and to this conclusion I come the rather when I see the diversities in the shapes of the eggs of different hens.

Harvey, William
Anatomical Exercises on the Generation of Animals
Exercise 59

Nothing so like as eggs; yet no one, on account of this appearing similarity, expects the same taste and relish in all of them.

Hume, David
An Enquiry Concerning Human Understanding
Section IV (p. 35)

The student of anatomy is perfectly well aware that there is not a single organ of the human body the structure of which does not vary, to a greater or less extent, in different individuals.

Huxley, Thomas H.
Man's Place in Nature
III (p. 166)

There is nothing stable in the world; uproar's your only music.

Keats, John
Letters of John Keats
Letter to George and Thomas Keats
13 January, 1818

. . . there are never in nature two beings which are exactly alike . . .

Leibniz, Gottfried Wilhelm
Leibniz: Discourse on Metaphysics
Monadology, 9

The starting point of Darwin's theory of evolution is precisely the existence of those differences between individual members of a race or species which morphologists for the most part rightly neglect. The first condition necessary, in order that any process of Natural Selection may begin among a race, or species, is the existence of differences among its members; and the first step in an enquiry into the possible effect of a selective process upon any character of a race must be an estimate of the frequency with which individuals, exhibiting any given degree of abnormality with respect to that character, occur. The unit, with which such an enquiry must deal, is not an individual but a race, or a statistically representative sample of a race; and the result must take the form of a numerical statement, showing the relative frequency with which the various kinds of individuals composing the race occur.

Pearson, Karl
Biometrika
Editorial, 1901 (p. 1)

Jucunda vicissitudo rerum.
[Variety is the spice of life.]

Proverb

Variation is, of course, an important characteristic of populations that individuals cannot have . . . A thousand exactly similar steel bearing balls (if such were possible) would be no more than one ball multiplied one thousand times. It is the quality of variation that makes it difficult at first to carry in mind a population in its complexity.

Tippett, L.C.
The World of Mathematics
Sampling and Standard Error
Volume 3 (p. 1480)

Sunt certi denique tines quos ultra citraque nequit existere verum.
[All variates are limited in both directions.]

Unknown

Variance is what any two statisticians are at.

Unknown

Since no two events are identical, every atom, molecule, organism, personality, and society is an emergent and, at least to some extent, a novelty.

Wheeler, William Maston
Proceedings of the Sixth International Congress of Philosophers
Emergent Evolution of the Social
Cambridge, Massachusetts

BIBLIOGRAPHY

Abbott, Edwin A. *Flatland*. Barnes & Noble, Inc., New York. 1963.

Abelson, Philip H. 'Editorial' in *Science*. 4 February 1994.

Adams, Douglas. *The Original Hitchhiker Radio Scripts*. Crown Publishers, Inc., New York. 1985.

Adams, Franklin. *Tobogganing on Parnassus*. Doubleday, Page & Company, Garden City. 1918.

Adams, Henry. *A Letter to American Teachers of History*. S.H. Furs & Co., Baltimore. 1910.

Adams, Henry. *The Education of Henry Adams*. Random House, New York. 1946.

Advertisement. *The American Statistician*. Volume 33, Number 4. November 1979.

Aeschylus. *The Plays of Aeschylus*. G. Bell, London. 1909.

Akenside, Mark. *The Poetical Works of Mark Akenside and John Dyer*. George Routledge, New York. 1855.

Alcott, Louisa May. *Little Women*. The World Publishing Co., Cleveland. 1946.

Allen, Arnold O. *Probability, Statistics, and Queueing Theory with Computer Science Applications* (Second Edition). Academic Press, Inc., Boston. 1990.

Allen, R.G.D. *Statistics for Economists*. Hutchinson & Co. Ltd., London. 1957.

Allibone, S. Austin. *Prose Quotations from Socrates to Macaulay*. J.B. Lippincott Co., Philadelphia. 1903.

Ambler, Eric. *A Coffin for Dimitrios*. The Sun Dial Press, New York. 1939.

Anderson, Poul. *New Scientist*. 25 September 1969.

Angell, Roger. *Late Innings: a Baseball Companion*. Simon & Schuster, New York. 1982.

Anscombe, F.J. 'Rejection of Outliers' in *Technometrics*. Volume 2. 1960.

Aquinas, Thomas. *Summa Theologiae*. McGraw-Hill Book Co., New York. 1975.

Arbuthnot, John. *Of the Laws of Chance*. Benj. Matte, London. 1692.

Arago. 'Eulogy on Laplace' in *Smithsonian Report*. 1874.

Aristotle. *Metaphysics*. Harvard University Press, Cambridge. 1947.

Aristotle. *On Generation and Corruption*. Translated by C.J.F. Williams. Oxford University Press, Oxford. 1982.

Aristotle. *On Interpretations*. Harvard University Press, Cambridge. 1962.

Aristotle. *On the Heavens (De caelo)*. Harvard University Press, Cambridge. 1939.

Aristotle. *The Poets*. Harvard University Press, Cambridge. 1939.

Aristotle. *The Physics (Physica)*. Harvard University Press, Cambridge. 1957.

Aristotle. *The Nicomachean Ethics (Ethica Nicomachea)*. Harvard University Press, Cambridge. 1939.

Aristotle. *The Art of Rhetoric (Rhetorica)*. Harvard University Press, Cambridge. 1959.

Arnauld, Antoine. *The Art of Thinking: Port-Royal Logic*. The Bobbs–Merrill Co., Inc., Indianapolis. 1964.

Arnold, Matthew. *Discourses in America*. Macmillan and Co., London. 1889.

Aron, Raymond. *The Opium of the Intellectuals*. Translated by Terence Kilmartin. Greenwood Press Publications, Westport. 1955.

Arthur, T.S. *Ten Nights in a Bar-Room and What I Saw There*. Edited by Donald A. Koch. The Belknap Press of Harvard University Press, Cambridge. 1964.

Asimov, Isaac. *Of Time and Space and Other Things*. Avon Books, New York. 1965.

Atherton, Gertrude. *Senator North*. John Wilson & Co., Cambridge. 1900.

Aurelius, Marcus. *The Meditations of the Emperor Antoninus Marcus Aurelius*. Translated by George Long. Thomas Y. Crowell & Co., New York. No date.

Austen, Jane. *Sense and Sensibility*. E.P. Dutton & Co., New York. 1908.

Bacon, Francis. *Advancement of Learning*. Ginn E. Company Publishers. No date.

Bacon, Francis. *New Atlantis*. D. Van Nostrand Co., Inc., New York. 1942.

Bacon, Francis. *The Novum Organon or A True Guide to the Interpretation of Nature*. The University Press, Oxford. 1855.

Baez, Joan. *Daybreak*. Dial Press, New York. 1968.

Bailey, Norman T.J. *The Mathematical Approach to Biology and Medicine*. John Wiley & Sons, London. 1967.

Bailey, Thomas D. 'Notable and Quotable' in *Wall Street Journal*. 17 December 1962.

Bailey, William B. and Cummings, John. *Statistics*. A.C. McClurg & Co., Chicago. 1917.

Balchin, Nigel. *The Small Back Room*. Collins, London. 1943.

Barrie, J.M. *The Greenwood Hat*. Charles Schribner's Sons, New York. 1938.

Barry, Frederick. *The Scientific Habit of Thought*. Columbia University Press, New York. 1927.

Bartlett, M.S. 'Discussion on Professor Pratt's Paper' in *Journal of the Royal Statistical Society*.

Bartlett, M.S. *Essays on Probability and Statistics*. Methuen & Co., Ltd., London. 1962.

Baudrillard, Jean. *Cool Memories*. Galilee, Paris. 1987.

Baughman, M. Dale. *Teacher's Treasury of Stories for Every Occasion*. Prentice-Hall, Inc., Englewood Cliffs. 1958.

Bell, Eric. *The Development of Mathematics*. McGraw-Hill Book Co., Inc., New York. 1940.

Bell, Eric. *Mathematics: Queen & Servant of Science*. McGraw-Hill Book Co., Inc., New York. 1951.

Belloc, Hilaire. *More Beasts (for Worse Children)*. E. Arnold, New York. 1897.

Belloc, Hilaire. *The Silence of the Sea*. Sheed & Wood, New York. 1940.

Bellow, Saul. *Herzog*. The Viking Press, New York. 1964.

Bellow, Saul. *The Dean's December*. Harper & Row, Publishers, New York. 1982.

Bennett, Arnold. *A Great Man*. George H. Dorian Co., New York. 1911.

Berdie, Douglas A. *Questionnaires: Design and Use*. The Scarecrow Press, Inc., Metuchen. 1974.

Bergson, Henri. *Creative Evolution*. Translated by Arthur Mitchell. Random House, Inc., Canada. 1944.

Berkeley, Edmund C. 'Right Answers—A Short Guide for Obtaining Them' in *Computers and Automation*. Volume 18, Number 10. September 1969.

Bernard, Claude. *An Introduction to the Study of Experimental Medicine*. Henry Schuman, Inc., New York 1949.

Bernard, Frederick R. *Printer's Ink*. Volume 138. 10 March 1927.

Beveridge, William Ian B. *The Art of Scientific Investigation*. Norton, New York. 1957.

Bierce, Ambrose. *The Devil's Dictionary*. Dover Publications, Inc., New York. 1958.

Billings, Josh. *Old Probability: Perhaps Rain—Perhaps Not*. Literature House, Upper Saddle River. 1970.

Blake, William. *The Complete Writings of William Blake; with variant readings*. Edited by Geoffrey Keynes. Oxford University Press, Oxford. 1966.

Bloch, Arthur. *Murphy's Law*. Price/Stern/Sloan Publishers, Inc., Los Angeles. 1979.

Blodgett, James H. 'Obstacles to Accurate Statistics' in *The Journal of the American Statistical Association*. New Series Number 41. March 1898.

Bohm, D. *Causality and Chance in Modern Physics*. D. van Nostrand Company, Inc., Princeton. 1957.

Boole, George. 'An Investigation of the Laws of Thought' in *Collected Logical Works*. Volume II. The Open Court Publishing Co., LaSalle. 1952.

Boorstin, Daniel J. *The Decline of Radicalism*. Vintage Books, New York. 1969.

Booth, Charles. *Charles Booth's London*. Penguin Books, Middlesex. 1971.

Borel, Emile. *Probability and Certainty*. Dover Publications, Inc., New York. 1964.

Borel, Emile. *Probabilities and Life*. Translated by Maurice Baudin. Dover Publications, New York. 1962.

Borges, Jorge Luis. *Ficciones*. Grove Press, Inc., New York. 1962.

Boring, E.G. 'When is Human Behavior Predetermined?' in *The Scientific Monthly*. Volume 84. 1957.

Born, Max. *Natural Philosophy of Cause and Chance*. Dover Publications, Inc., New York. 1964.

Bostwick, Arthur E. 'The Theory of Probabilities' in *Science*. Volume III, Number 54. 10 January 1896.

Boswell, James. *The Life of Samuel Johnson*. E.P. Dutton Co., New York. No date.

Boudreau, Frank G., MD and Kiser, Clyde V. *Problems in the Collection and Comparability of International Statistics*. Milbank Memorial Fund, New York. 1949.

Boulle, Pierre. *The Bridge Over the River Kwai*. Translated by Xan Fielding. The Vanguard Press, Inc., New York. 1954.

Bowley, Arthur L. *Elements of Statistics*. Staples Press Limited, London. 1946.

Bowley, Arthur L. *The Mathematical Gazette*. Volume 12, Number 77, July 1925.

Bowman, Scotty. 'A Lot More Where They Come From' in *Sports Illustrated*. 2 April 1973.

Box, G.E.P. 'Use and Abuse of Regression' in *Technometrics*. Volume 8, Number 4. November 1966.

Box, G.E.P. 'Discussion' in *Journal of the Royal Statistical Society*. Series B., 18. 1956.

Bradley, F.H. *The Principles of Logic*. Oxford University Press, London. 1883.

Braude, Jacob M. *Complete Speaker's and Toastmaster's Library*. Prentice-Hall, Inc., Englewoods Cliffs. 1967.

Browning, Elizabeth Barrett. *The Complete Poetical Works of E.B.B.* Houghton, Mifflin & Co., New York. 1900.

Browning, Robert. *The Poems and Plays of Robert Browning*. Random House, Inc., New York. 1961.

Browning, Robert. *The Ring and the Book*. Oxford University Press, London. 1912.

Bruner, Jerome S. *The Process of Education*. Harvard University Press, Cambridge. 1965.

Buchner, Ludwig. *Force and Matter*. Truth Seeker Company, New York. 1950.

Bulwer, Lytton E.G. *Eugene Aram*. Collier, New York. 1901.

Burgess, Robert W. 'The Whole Duty of the Statistical Forecaster' in *Journal of the American Statistical Association*. Volume 32, Number 200. December 1937.

Burnan, Tom. *The Dictionary of Misinformation*. Thomas Y. Crowell Co., New York. 1975.

Burney, Frances. *Camilla*. Oxford University Press, London. 1972.

Burns, Robert. *The Complete Poetical Works of Robert Burns*. Houghton, Mifflin & Co., Boston. 1897.

Butler, Joseph. *The Analogy of Religion*. J.M. Dent & Co., London. 1906.

Butler, Samuel. *Erewhon or Over the Range*. University of Delaware Press, Newark. 1981.

Butler, Samuel. *Hudibras*. The Clarendon Press, Oxford. 1967.

Butler, Samuel. *Samuel Butler's Note-Books*. Selections edited by Geoffrey Keynes and Brian Hill. Jonathan Cape, London. 1951.

Butler, Samuel. *The Poetical Works*. Volume II. Bell and Daldy, York Street, London. 1854.

Byron, Lord. *Don Juan*. Edited by Truman Guy Steffon and Willis W. Pratt. University of Texas Press, Austin. 1957.

Byron, Lord. *The Complete Poetical Works of Byron*. Houghton Mifflin & Co., Boston. 1933.

Cage, John. *Silence 1961*. Wesleyan University Press, Middletown. 1961.

Cahier, Charles. *Quelques Six Mille Proverbes*. Julien, Lanier et cie, Paris. 1856.

Camus, Albert. *The Fall*. Vintage Books, New York. 1956.

Cardozo, Benjamin N. 'Mr. Justice Holmes' in *Harvard Law Review*. Volume 44, March 1931.

Cardozo, Benjamin. *The Growth of Law*. Yale University Press, New Haven. 1924.

Carlyle, Thomas. *English and other Critical Essays*. Dutton, New York. 1964.

Carlyle, Thomas. *Sartor Resartus*. C. Scriber's Sons, New York. 1921.

Carrel, Alexis. *Man The Unknown*. Harper & Brothers Publishers, New York. 1939.

Carroll, Lewis. *The Complete Works of Lewis Carroll*. The Modern Library, New York. 1936.

Carroll, Lewis. *Alice's Adventures in Wonderland*. Heirloom Library, London. 1949.

Cervantes, Miguel de. *The Ingenious Gentleman Don Quixote de la Mancha*. The Viking Press, New York. 1949.

Chambers, Robert. *Vestiges of the Natural History of Creation, with extensive additions and emendations* (Second Edition). J. Churchill, London. 1844.

Chamfort, Sebastien Roch. *Maximes et pensèes.* Le livre de poche, Paris. 1970.

Champernowne, D.G. *Journal of the Royal Statistical Society.* Volume 118. 1955.

Chappell, Edwin. *The Tangier Papers of Samuel Pepys.* Printed for the Navy Records Society, London. 1935.

Chaucer, Geoffrey. *Troylus & Cryseyde.* Centaur Press, Ltd, London. No date.

Chernoff, Herman and Moses, Lincoln E. *Elementary Decision Theory.* John Wiley & Sons, Inc., New York. 1959.

Chesterson, G.K. *The Father Brown Omnibus.* Dodd, Mead & Co., Inc., New York. 1951.

Chesterson, G.K. *The Wisdom of Father Brown.* Dodd, Mead & Co., New York. 1930.

Chestov, Leon. 'Look Back and Struggle' in *Forum Philosocum.* Volume 1, Number 1. 1930.

Churchill, Winston S. *The Story of the Malakand Field Force.* W.W. Norton & Co., New York. 1990.

Cicero. *Cicero: De Senectute, De Amicitia, De Divinatione.* Translated by William Armistead Falconer. Harvard University Press, Cambridge. 1959.

Cicero. *Epistolae ad atticum.* Belles Lettres, Paris. 1984.

Cicero. *Orationes Philippicae.* University of North Carolina Press, Chapel Hill. 1985.

Clark, Arthur C. *The Lost Worlds of 2001.* Gregg Press, Boston. 1979.

Clark, Ronald W. *Einstein: The Life and Times.* The World Publishing Co., New York. 1971.

Coats, R.H. 'Science and Society' in *Journal of the American Statistical Society.* Volume 34, Number 205. March 1939.

Coates, Robert M. 'The Law' in *The World of Mathematics, Volume IV* by James R. Newman. Simon and Schuster, New York. 1956.

Cochran, William G. and Cox, Gertrude M. *Experimental Designs.* John Wiley & Sons, Inc., New York. 1957.

Cochran, William G. *Sampling Techniques.* John Wiley & Sons Co., New York. 1977.

Cohen, Jacob. *Statistical Power Analysis of the Behavioral Sciences* (Second Edition). Lawrence Erlbaum Associates, Publishers, New Jersey. 1988.

Cohen, Jerome. 'Tense Triangle—What to Do About Taiwan' in *Time.* 7 June 1971.

Cohen, John. *Chance, Skill, and Luck.* Penguin Books, Baltimore. 1960.

Cohen, Morris R. 'The Statistical View of Nature' in *Journal of the American Statistical Association.* Volume 31, Number 194. June 1936.

Cohen, Morris R. *A Preface to Logic.* Meridian Books, New York. 1944.

Colton, Charles. *Lacon or Many Things in A Few Words*. William Gowan, New York. 1849.

Comfort, Alex. *Darwin and the Naked Lady: Discursive Essays on Biology and Art*. George Braziller, New York. 1962.

Conrad, Joseph. *Lord Jim*. Doubleday, Doran and Co., Inc. for Wm. H. Wise & Co., Garden City. 1928.

Cook, Robin. *Mortal Fear*. G.P. Putnam's Sons, New York. 1988.

Coole, W.P. 'Letters to the Editor' in *The American Statistician*. Volume 23, Number 1. February 1969.

Cooley, Charles. 'Observations on the Measure of Change' in *Journal of the American Statistical Association*. New Series, Number 21. March 1893.

Copernicus, Nicolaus. *On the Revolutions of the Heavenly Spheres*. Translated by Charles Glenn Wallis. Encyclopaedia Britannica, Inc. 1939.

Cort, David. *Social Astonishments*. The Macmillan Co., New York. 1963.

Cowper, William. *Cowper: Poetical Works*. Oxford University Press, London. 1967.

Cox, D.R. and Hinkley, D.V. *Theoretical Statistics*. Chapman and Hall, London. 1974.

Crawford, Marion F. *Don Orsino*. Macmillan, New York. 1914.

Crichton, Michael. *Rising Sun*. Alfred A. Knopf, New York. 1992.

Crichton, Michael. *Sphere*. Ballantine Books, New York. 1987.

Crichton, Michael. *The Terminal Man*. Alfred A. Knopf, New York. 1974.

Crick, Francis. *Life Itself, Its Original Nature*. Simon Schuster, New York. 1981.

Cronbach, L.J. 'The Two Disciplines of Scientific Psychology' in *The American Psychologist*. Volume 12. November 1957.

Crothers, Samuel McChord. *The Gentle Reader*. Books for Libraries Press, Freeport. 1972.

da Vinci, Leonardo. *The Notebooks of Leonardo Da Vinci*. Edited by Edward MacCurdy. George Braziller, New York. 1939.

Dampier-Whetham, William. *A History of Science: Its Relation with Philosophy and Religion*. Cambridge University Press, London. 1930.

Dampier-Whetham, William. *Science and the Human Mind*. Longman, Green, and Co., London. 1912.

Dante, Alighieri. *The Divine Comedy of Dante Alighieri*. Harvard University Press, Cambridge. 1918.

Darwin, Charles. *The Life and Letters of Charles Darwin*. Volume I. D. Appleton and Co., New York. 1888.

Darwin, Charles. *The Life and Letters of Charles Darwin*. Volume II. D. Appleton and Co., New York. 1888.

Darwin, Charles. *The Origin of Species by Means of Natural Selection*. D. Appleton, New York. 1896.

Davies, J.T. *The Scientific Approach*. Academic Press, New York. 1973.

Davies, Robertson. *The Diary of Samuel Marchbanks*. Clark, Irwin & Co., Ltd, Toronto. 1947.

Davis, Joseph S. 'Statistics and Social Engineering' in *Journal of the American Statistical Association*. Volume 32, Number 197. March 1937.

Dawkins, Richard. *The Blind Watchmaker*. Norton, New York. 1986.

de Finetti, B. *Theory of Probability*. Wiley, Chichester. 1974.

de Jonnes, Moreau. *Éléments de Statistique* (Second Edition). Paris. 1856.

de Jouvenel, Bertrand. *The Art of Conjecture*. Basic Books, New York. 1967.

de Leeuw, A.L. *Rambling through Science*. Whittlesey House, New York. 1932.

de Moivre, A. *The Doctrine of Chances*. Chelsea Publishing Co., New York. 1967.

de Morgan A. *A Budget of Paradoxes*. The Open Court Publishing Co., Chicago. 1915.

de Solla Price, Derek John. *Little Science, Big Science*. Columbia University Press, New York. 1986.

de Spinoza, Benedict. *Ethics*. J.M. Dent & Sons, Ltd., London. 1941.

De Vries, Peter. *Ruben, Ruben*. Little, Brown and Co., Boston. 1964.

Deming, William Edwards. *Out of the Crisis*. Massachusetts Institute of Technology, Cambridge. 1991.

Deming, William Edwards. *Sample Design in Business Research*. John Wiley & Sons, Inc., New York. 1960.

Deming, William Edwards. *Some Theory of Sampling*. John Wiley & Sons, Inc., New York. 1950.

Deming, William Edwards. *Statistical Adjustment of Data*. J. Wiley & Sons, Inc., New York. 1943.

Deming, William Edwards. 'Some Principles on the Shewhart Methods of Quality Control' in *Mechanical Engineering*. Volume 66. March 1944.

Deming, William Edwards. 'On the Classification of Statistics' in *The American Statistician*. Volume 2, Number 2. April 1948.

Deming, William Edwards. 'On the Presentation of the Results of Sample Surveys as Legal Evidence' in *Journal of the American Statistical Association*. Volume 49, Number 268, December 1954.

Deming, William Edwards. 'On a Classification of the Problems of Statistical Inference' in *Journal of the American Statistical Association*. Volume 37, Number 218. June 1942.

Descartes, René. *Discourse on the Method of Rightly Conducting the Reason and Seeking Truth in Sciences*. The Open Court Publishing Co., Inc., Chicago. 1907.

Descartes, René. *Rules for the Direction of the Mind*. The Bobbs–Merrill Co., Inc., Indianapolis. 1961.

Deutscher, I. 'Public and Private Opinions: Social Situations and Multiple Realities' in *The Social Contexts of Research*. Edited by S.Z. Nagi and R.G. Corwin. Wiley, London. 1972.

Devons, Ely. *Essay on Economics*. Greenwood Press, Westport. 1961.

Dewey, John. *Logic: The Theory of Inquiry*. Irvington Publishers, Inc., New York. 1982.

Dewey, John. *Art as Experience*. George Allen & Unwin, Ltd., London. 1934.

Dickens, Charles. *The Work of Charles Dickens*. Charles Scribner's Sons, New York. 1902.

Dickson, Paul. *The Official Rules*. Dalcorte Press, New York. 1978.

Disney, Dorothy. *Crimson Friday*. Dell, New York. 1946.

Disraeli, Benjamin. *Sybil or The Two Nations*. T.A. Contall, Ltd., Edinburgh. 1927.

Doyle, Sir Arthur Conan. *The Complete Sherlock Holmes*. Doubleday & Co., Inc., Garden City. 1927.

Driscoll, Michael F. 'The Ten Commandments of Statistical Inference' in *The American Mathematical Monthly*. Volume 84, Number 8. 1977.

Dryden, John. *The Poetical Works of Dryden*. Edited by George R. Noyse. Houghton Mifflin Co., Boston. 1950.

Dubos, René. *Louis Pasteur: Free Lance of Science*. Charles Scribner's Sons, New York. 1976.

Duckworth, George E. *The Complete Roman Drama, Volume Two*. Random House, New York. 1966.

Durand, David. 'A Dictionary for Statismagicians' in *The American Statistician*. Volume 24, Number 3. June 1970.

Eco, Umberto. *Il pendolo di Foucault*. Translated by William Weaver. Harcourt Brace Jovanovich, San Diego. 1989.

Eddington, A.S. *New Pathways in Science*. Cambridge University Press, Cambridge. 1935.

Eddington, A.S. *The Nature of the Physical World*. The Macmillan Co., New York. 1930.

Eddington, A.S. *Space, Time and Gravitation*. Cambridge University Press, Cambridge. 1920.

Edge, David O. and Mulkay, Michael J. *Astronomy Transformed*. John Wiley & Sons, New York. 1976.

Edgeworth, Francis Ysidro. 'The Philosophy of Chance' in *Mind*. Volume 9. 1884.

Edgeworth, Francis Ysidro. 'On the Use of the Theory of Probabilities in Statistics Relating to Society' in *Journal of the Royal Statistical Society*. January 1913.

Edgeworth, Francis Ysidro. *Journal of the Royal Statistical Society*. Volume 53. 1890.

Edgeworth, Francis Ysidro. 'On the Representation of Statistics by Mathematical Formula (concluded)' in *Journal of the Royal Statistical Society*. Volume XLII. 1899.

Edwards, A.W.F. *Likelihood*. Cambridge University Press, Cambridge. 1972.

Ehrenberg, A.S.C. *Data Reduction*. John Wiley & Sons, New York. 1975.

Einstein, Albert. *Sidelights on Relativity*. Methuen & Co., Ltd., London. 1922.

Einstein, Albert and Infield, Leopold. *The Evolution of Physics*. Simon and Schuster, New York. 1938.

Eisenhart, Churchill. 'The Role of a Statistical Consultant in a Research Organization' in *The American Statistician*. Volume 2, Number 2. April 1948.

Eldridge, Paul. *Maxims for a Modern Man*. Thomas Yoseloff: Publisher, New York. 1965.

Eliot, George. *Daniel Deronda*. Volume 1. J.M. Dent & Sons Ltd., London. 1964.

Eliot, George. *The George Eliot Letters*. Edited by Gordon S. Haight. Volume II. Yale University Press, New Haven. 1954.

Eliot, George. *The George Eliot Letters*. Edited by Gordon S. Haight. Volume IV. Yale University Press, New Haven. 1954.

Eliot, George. *The Mill on the Floss*. The Clarendon Press, Oxford. 1980.

Eliot, George. 'Impressions of Theophrastus Such' in *The Complete Works of George Eliot*. The Kelmscott Society Publisher, New York. No date.

Ellis, Havelock. *The Dance of Life*. Houghton Mifflin Co., Boston. 1923.

Ellison, Harlon. *Dangerous Visions*. Berkley Books Publishing Group, New York. 1967.

Emerson, Ralph Waldo. *Essays*. 1st Series. Houghton Mifflin & Co., Boston. 1883.

Emerson, Ralph Waldo. *Essays*. 2nd Series. Houghton Mifflin & Co., Boston. 1886.

Emerson, Ralph Waldo. *Lectures and Biographical Sketches*. Houghton, Mifflin & Co., New York. 1884.

Emerson, Ralph Waldo. *The Conduct of Life*. Ticknor and Fields, Boston. 1861.

Emerson, Ralph Waldo. *The Journals of Ralph Waldo Emerson*. Random House, Inc., New York. 1960.

Esar, Evan. *Esar's Comic Dictionary*. Bantam Doubleday Dell Publishing Group, Inc., New York. 1943.

Ettorre, Barbara. *Harper's Magazine*. Volume 249, Number 1491. August 1974.

Euripides. *The Plays of Euripides*. Translated by Shelly Dean Milman. E.P. Dutton & Co., New York. 1906.

Evans, Bergen. *The Natural History of Nonsense*. A.A. Knopf, New York. 1946.

Fabing, Howard and Marr, Ray. *Fischerisms*. The Science Press Printing Co., Lancaster. 1937.

Farr, William. See Marion Diamond and Mervyn Stone, 'Nightingale on Quetelet' in *Journal of the Royal Statistical Society*. Series A, Number

144. 1981.

Feller, William. *An Introduction to Probability Theory and Its Applications*. Volume 1. John Wiley & Sons, Inc., New York. 1960.

Feynman, Richard P., Leighton, Robert B. and Sands, Matthew. *The Feynman Lectures on Physics*. Volume I. Addison-Wesley Publishing Co., Reading. 1963.

Fiedler, Edgar R. 'The Three Rs of Economic Forecasting—Irrational, Irrelevant and Irreverent' in *Across the Board*. June 1977.

Fienberg, Stephen E. 'Graphical Methods in Statistics' in *The American Statistician*. Volume 13, Number 4. November 1979.

Finney, D.J. 'The Questioning Statistician' in *Statistics in Medicine*. Volume I. 1982.

Fischer, Robert B. *Science, Man and Society*. W.B. Saunders Co., Philadelphia. 1971.

Fisher, Sir Ronald A. *Statistical Methods for Research Workers*. Hafner Press, New York. 1970.

Fisher, Sir Ronald A. *The Design of Experiments*. Oliver and Boyd, Edinburgh. 1937.

Fisher, Sir Ronald A. 'The Expansion of Statistics' in *American Scientist Magazine*. Volume 42, Number 2. April 1954.

Fisher, Sir Ronald A. 'Student' in *Annals of Eugenics*. Volume 9. 1939.

Fisher, R.A. 'The Design of Field Experiments' in *Journal of the Ministry of Agriculture of Great Britain*. Volume 33. 1926.

Fisher, Sir Ronald A. 'Statistical Methods and Scientific Induction' in *Journal of the Royal Statistical Society*. Series B, Number 17. 1955.

Fisher, Sir Ronald A. *Sankya*. Volume 4. 1938.

Fitzgerald, F. Scott. *This Side of Paradise*. Charles Scribner's Sons, New York. 1948.

Fleiss, Joseph L. 'Letters to the Editor' in *The American Statistician*. Volume 21, Number 4. October 1967.

Flesch, Rudolf. *The New Book of Unusual Quotations*. Harper & Row, New York. 1966.

Forbes, J.D. 'On the Alleged Evidence for a Physical Connection between Stars Forming Binary or Multiple Groups, Deduced from the Doctrine of Chances' in *The London, Edinburgh and Dublin Philosophical Magazine and Journal of Science*. Volume 37. December, 1850.

Forster, E.M. *Howards End*. Holmes & Meier Publishers, Inc. New York. 1973.

Foss, Sam Walter. *Back Country Poems*. Lee & Shepard, Boston. 1894.

Fourier, Jean Baptiste Joseph. *Analytical Theory of Heat*. Translated, with notes, by Alexander Freeman. The University Press, Cambridge. 1878.

Fox, Russell, Garbuny, Max and Robert Hooke. *The Science of Science*. Walker and Co., New York. 1963.

Freeman, Linton C. *Elementary Applied Statistics in Behavioral Science*. John Wiley & Sons Co., New York. 1965.

Freeman, R. Austin. *A Certain Doctor Thorndyke*. Hodder & Stoughton, London. 1944.

Freidman, Martin. 'Irresponsible Monetary Policy' in *Newsweek*. 10 January 1972.

Freud, Sigmund. 'On Narcissism' in *Collected Papers*. Volume II. Translated by Cecil M. Baines. The Hogarth Press, London.

Friedman, Thomas L. *From Beirut to Jerusalem*. Anchor Books, New York. 1989.

Froude, James Anthony. *Short Studies on Great Subjects*. Charles Scribner's Sons, New York. 1892.

Fry, Thornton C. *Probability and Its Engineering Uses*. D. Van Nostrand Co., Princeton. 1965.

Galsworthy, John. *End of the Chapter*. Charles Scribner's Sons, New York. 1937.

Galton, Francis. *Hereditary Genius: An Inquiry into Its Laws and Consequences*. Macmillan and Co., Ltd., London. 1914.

Galton, Francis. *Inquiries into Human Faculty and Its Development*. J.M. Dent & Co., London. 1908.

Galton, Francis. *Memories of My Life*. Methuen & Co., London. 1908.

Galton, Francis. *Natural Inheritance*. Macmillan and Co., New York. 1889.

Galton, Francis. 'Kinship and Correlation' in *North American Review*. Volume 150. 1890.

Galton, Francis. 'Co-relations and Their Measurements, Chiefly from Anthropometric Data' in *Proceedings of the Royal Society of London*. Volume 45. 1888.

Gann, Ernest K. *Brain 2000*. Doubleday & Co., Inc., Garden City. 1980.

Garson, Barbara. *MacBird*. Grassy Knoll Press, Berkeley. 1966.

Gay, John. *John Gay: Poetry and Prose*. Volume II. The Clarendon Press, Oxford. 1974.

Geary, R.C. 'Testing for Normality' in *Biometrika*. Volume 34. 1947.

Gibbon, Edward. *Autobiography*. E.P. Dutton, New York. 1932.

Gibbon, Edward. *The Decline of the Roman Empire*. Modern Library, New York. 1932.

Gilbert, W.S. and Sullivan, Arthur. *The Complete Plays of Gilbert and Sullivan*. Garden City Publishing Co., Inc., Garden City. 1938.

Gilbert, William. *On the Loadstone and Magnetic Bodies and on the Great Magnet the Earth*. Translated by P. Fleury Mottelay. Edwards Brothers, Inc., Ann Arbor. 1941.

Gilman, Charlotte. *Human Work*. McClure, Phillips & Co., New York. 1904.

Ginsberg, Allen. *America*. The Coach House Press, Toronto. 1972.

Gissing, George. *New Grub Street*. The Modern Library, New York. 1926.

Glantz, S.A. 'Biostatistics: How to Detect, Correct and Prevent Errors in the Medical Literature' in *Circulation*. Volume 61. 1980.

Gleick, James. *Chaos*. Viking Penguin, Inc., New York. 1987.

Godwin, William. *St. Leon: A Tale of the Sixteenth Century*. McGrath Publishing Co., New York. 1972.

Goldwyn, Samuel. 'Obituary' in *New York Times*. 1 February 1974.

Good, I.J. 'Kinds of Probability' in *Science*. Volume 129. 20 February 1959.

Goodman, Richard. *Modern Statistics*. ARC Books, Inc., New York. 1964.

Greedman, D.A. and W.C Navidi. 'Regression Models for Adjusting the 1980 Census' in *Statistical Science*, Volume 1, Number 1. 1986.

Green, Celia. *The Decline and Fall of Science*. Hamilton, London. 1976.

Greenwood, M. 'Discussion to the paper Some Aspects of the Teaching of Statistics' in *Journal of the Royal Statistical Society*. Volume 102. 1939.

Greer, Scott. *The Logic of Social Inquiry*. Aldine Publishing Co., Chicago. 1973.

Guest, Judith. *Ordinary People*. The Viking Press, New York. 1976.

Guillen, Michael. *Bridges to Infinity*. Jeremy P. Tarcher, Inc., Los Angeles. 1983.

Gunther, John. *Taken at the Flood: The Story of Albert D. Lasker*. Harper & Brothers Publishers, New York. 1960.

Habera, Audrey and Runyon, Richard P. *General Statistics*. Addison-Wesley Publishing Co., Inc., Reading. 1973.

Hacking, Ian. *The Emergence of Probability*. Cambridge University Press, Cambridge. 1975.

Hailey, Arthur. *Airport*. Doubleday, Garden City. 1968.

Haldeman, H.R. *The Ends of Power*. Times Books, New York. 1978.

Hamilton, Edith. *The Roman Way*. W.W. Norton & Co., Inc., New York. 1960.

Hamming, Richard. *The Art of Probability for Scientists and Engineers*. Addison-Wesley Publishing Co., Redwood City.

Hammond, Kenneth and Adelman, Leonard. 'Science, Values, and Human Judgment' in *Science*. Volume 194, Number 4263. 22 October 1976.

Hancock, William Keith. *Australia*. London. 1930.

Hand, D.J. 'The Role of Statistics in Psychiatry' in *Psychological Medicine*. Volume 15, 1985.

Hardy, Thomas. *Tess of the d'Urbervilles*. A.L. Burt Co., New York. 1919.

Harris, Errol E. *Hypothesis and Perception*. George Allen & Unwin Ltd., London. 1970.

Harrison, Harry. *Astounding*. Random House, New York. 1973.

Harte, Francis Bret. *Two Men of Sandy Bar*. Collier & Son, New York. 1904.

Hayford, F. Leslie. 'Some Uses of Statistics in Executive Control' in *Journal of the American Statistical Association*. Volume 31, Number 193. March 1936.

Heaviside, Oliver. *Electromagnetic Theory*. D. Van Nostrand, New York. 1893.

Heinlein, Robert A. *Time Enough for Love*. G.P. Putnam's Sons, New York. 1973.

Heinlein, Robert A. *To Sail Beyond the Sunset*. G.P. Putnam's Sons, New York. 1987.

Heise, David R. *Causal Analysis*. John Wiley & Sons, New York. 1975.

Heisenberg, Werner. *The Physical Principles of the Quantum Theory*. Dover Publications, Inc., New York. 1930.

Helvetius, C.A. *On Mind*. Burt Franklin, New York. 1970.

Henry, O. *Tales of O. Henry*. Doubleday & Co., Inc., Garden City. 1969.

Herbert, Nick. *Quantum Reality*. Anchor Press, Garden City. 1985.

Herodotus. *The History of Herodotus*. Volume II. E.P. Dutton & Co., Inc., New York. No date.

Herschel, John F.W. *Outlines of Astronomy*. Longman & Green, London. 1828.

Heyward, DuBose. *Carolina Chansons*. Macmillan Publishing Co., New York. 1922.

Heyworth, Sir Geoffrey. 'Inaugural Address' in *Journal of the Royal Statistical Society*. Volume 113, Number 4. 1950.

Hobbes, Thomas. *Leviathan*. E.P. Dutton & Co., Inc., New York. No date.

Hoel, Paul G. *Introduction to Mathematical Statistics* (Third Edition). John Wiley & Sons, Inc., New York. 1962.

Hoffer, Eric. *The True Believer: thoughts on the nature of man's movement*. Perennial Library, New York. 1951.

Hogben, L. *Statistical Theory*. Allen and Unwin, London. 1957.

Hogben, Lancelot. *Science in Authority*. Unwin University Books, London. 1963.

Holmes, O.W. *The Complete Poetical Works of Oliver Wendell Holmes*. Houghton, Mifflin and Co., Boston. 1881.

Holmes, O.W. *Pages from an Old Volume of Life*. Houghton, Mifflin and Co., Boston. 1890.

Holmes, O.W. *The Professor at the Breakfast Table*. J.C. Holten, London. 1869.

Holmes, O.W. *The Autocrat of the Breakfast Table*. Houghton, Mifflin and Co., Boston. 1894.

Holmes, O.W., Jr. *Collected Legal Papers*. Harcourt, Brace and Howe, Inc. 1920.

Holmes, O.W., Jr. 'Natural Law' in *Harvard Law Review*. Volume 82. 1918.

Holmes, O.W., Jr. 'Path of the Law' in *Harvard Law Review*. Volume 10. 1897.

Homer. *The Iliad of Homer*. Translated by H. Hailstone. Relfe Brothers, London. 1882.

Hood, Thomas. *Miss Kilmansegg & Her Precious Legs; A Golden Legend*. E. Moxon & Sons, London. 1870.

Hopkins, Harry. *The Numbers Game: The Bland Totalitarianism*. Martin Secker & Warburg Ltd, London. 1973.

Horace. 'The Golden Mean' in *The Complete Works of Horace*. Translated by Herbert Wetmore Wells. Random House, Inc., New York. 1936.

Horace. *The Satires and Epistles of Horace*. Translated by Smith Palmer Bovie. The University of Chicago Press, Chicago. 1959.

Howe, E.W. *Sinner Sermons*. Halderman–Julius Co., Girard. 1926.

Howitt, Mary. *The Poems of Mary Howitt*. Hurst & Co., Publishers, New York.

Hoyle, F. *Galaxies, Nuclei, and Quasars*. Heinemann, London. 1965.

Hubbard, Elbert. *The Philistine: A Periodical of Protest*. Volume XI. Society of Philistine, East Aurora. July 1900.

Hubble, Edwin. *The Nature of Science and other Lectures*. Greenwood Press, Publishers, Westport. 1977.

Huff, Darrell. *How to Lie with Statistics*. W.W. Norton & Co., Inc., New York. 1954.

Hugo, Victor. *Les Misérables*. Translated by Isabel F. Hapgood. T.Y. Crowell & Co., New York. 1887.

Hume, David. *An Enquiry Concerning Human Understanding*. The Open Court Publishing Co., Chicago. 1921.

Hume, David. *A Treatise of Human Nature*. Penguin Books, Baltimore. 1969.

Hunter, Evan. *The Paper Dragon*. Dell Publishing Co., Inc. 1966.

Huxley, Aldous. *Brave New World*. Harper & Row, Publishers, New York. 1946.

Huxley, Aldous. *Literature and Science*. Harper & Row, Publishers, New York. 1963.

Huxley, Aldous. *Proper Studies*. Chatto & Windus, London. 1949.

Huxley, Aldous. *Stories, Essays, and Poems*. J.M. Dent & Sons, Ltd., London. 1949.

Huxley, Aldous. *Time Must Have a Stop*. Harper & Brothers Publishers, New York. 1944.

Huxley, Thomas H. *Man's Place in Nature*. The University of Michigan Press, Ann Arbor. 1959.

Huxley, Thomas. *Method and Results*. D. Appleton and Co., New York. 1896.

Huxley, Thomas. *Collected Essays*. 'On Descartes' "Discourse Touching the Method of Using One's Reason Rightly and of Seeking Scientific Truth"'. Volume I. Macmillan and Co., Limited. London 1904.

Huxley, Thomas. *Collected Essays*, 'Biogenesis and Abiogenesis'. Volume VIII. Macmillan and Co., Limited. London 1908.

Huygens, Christiaan. *Treatise on Light*. Rendered into English by Silvanus P. Thompson. Macmillan & Co., London. 1912.

Inge, William Ralph. *Outspoken Essays*. Longman Green Co., London. 1920.

Jacobs, Joseph. 'The Middle American' in *American Magazine*. Volume 63, March 1907.

Jahoda, Marie, Morton, Deutsch and Cook, Stuart W. *Research Methods in Social Relations*. Volume 1. Dryden Press, New York. 1951.

James, Henry. *The Spoils of Poynton*. W. Heinemann, London. 1897.

James, P.D. *Death of an Expert Witness*. Charles Scribner's Sons, New York. 1977.

James, William. *The Principles of Psychology*. Dover Publications, New York. 1918.

James, William. 'The Dilemma of Determinism' in *Unitarian Review and Religious Magazine*. Volume XXII, Number 3. September, 1884.

Jeans, James. *Physics and Philosophy*. Dover Publications, Inc., New York. 1981.

Jeans, J.H. *The New Background of Science*. Cambridge University Press, Cambridge. 1934.

Jefferys, Harold. 'Probability and Scientific Method' in *Proceedings of the Royal Statistical Society*, Series A, Volume 146. 1934.

Jevons, W.S. *The Principles of Science*. Macmillan and Co., New York. 1887.

Johnson, Palmer O. 'Modern Statistical Science and its Function in Educational and Psychological Research' in *The Scientific Monthly*. June 1951.

Johnson, Samuel. 'The Idler and the Adventurer' in *The Yale Edition of the Works of Samuel Johnson*. Yale University Press, New Haven. 1958.

Johnston, Alva. *The Legendary Mizners*. Farrar, Strauss & Young, New York. 1953.

Jones, Raymond F. *The Non-Statistical Man*. Belmont Productions, Inc., New York. 1964.

Jonson, Ben. *Volpone*. Chandler Publishing Co., San Francisco. 1961.

Juster, Norton. *The Dot and the Line; a romance in lower mathematics*. Random House, New York. 1963.

Juster, Norton. *The Phantom Tollbooth*. Epstein & Carroll Associates, Inc., New York. 1962.

Kac, Mark. *Probability and Related Topics in Physical Science*. Interscience Publishers, Inc., New York. 1959.

Kac, Mark. 'Statistical Independence in Probability Analysis and Number Theory' in *The Carus Mathematical Monograph*, Number Twelve. The Mathematical Association of America. 1959.

Kadane, Joseph. 'Comment' in *Statistical Science*. Volume 1, Number 1, February 1986.

Kadanoff, Leo P. 'Complete Structure from Simple Systems' in *Physics Today*. March 1991.

Kant, Immanuel. 'The Critique of Judgment' in *Philosophical Writings*. Edited by Ernst Behler. Continuum, New York. 1986.

Kapitza, Peter Leonidovich. 'Science East and West: Reflections of Peter Kapitza' in *Nature*. Volume 288. 11 December 1980.

Kaplan, Abraham. *The Conduct of Inquiry*. Chandler Publishing Co., San Francisco. 1964.

Karpansky, L. *High School Education*. New York. 1912.

Kasner, Edward and Newman, James. *Mathematics and the Imagination*. Simon and Schuster, New York. 1967.

Keats, John. *Letters of John Keats*. With an introduction by Hugh l'Aanson Fausset. Thomas Nelson and Sons, Ltd., London. No date.

Keegan, John. *The Face of Battle*. The Viking Press, New York. 1976.

Keeney, Ralph L. and Raiffa, Howard. *Decisions with Multiple Objectives: Preferences and Value Tradeoffs*. John Wiley & Sons, New York. 1976.

Kelly-Bootle, Stan. *The Devil's DP Dictionary*. McGraw-Hill Book Co., Inc., New York. 1981.

Kendall, M.G. 'The History and Future of Statistics' in *Statistical Papers in Honor of George Snedecor*. Edited by T.A. Bancroft. Iowa State University Press, Ames. 1972.

Kendall, M.G. and Stuart, A. *The Advanced Theory of Statistics*. Volume I. C. Griffin, London. 1947.

Kendall, Maurice G. 'Hiawatha Designs an Experiment' in *The American Statistician*. Volume 13, Number 5. December 1959.

Kendall, Maurice G. 'Who Discovered the Latin Square?' in *The American Statistician*. Volume 11, Number 4. August 1948.

Kerridge, D.F. 'Discussion on Paper by Dr. Marshall and Professor Olkin' in *Journal of the Royal Statistical Society*. Series B, Volume 30. 1968.

Keynes, John Maynard. *A Treatise on Probability*. Macmillan and Co., Limited, London. 1979.

King, Willford. 'Consolidating Our Gains' in *Journal of the American Statistical Association*. Volume 31, Number 193. March 1936.

Kipling, Rudyard. *From Sea to Sea*. Doubleday, Page & Co., Garden City. 1913.

Kipling, Rudyard. *Rudyard Kipling's Verse*. Doubleday and Co., Inc., New York. 1940.

Klagsbrun, Francine. *The First Ms Reader*. Warner Paperback Library, New York. 1973.

Kneale, W.C. *Probability and Induction*. The Clarendon Press, Oxford. 1952.

Knebel, Fletcher. *Reader's Digest*. December 1961.

Kolmogorov, A.N. *Foundations of the Theory of Probability*. Chelsea Publishing Co., New York. 1956.

Koshland, Daniel E., Jr. 'Editorial' in *Science*. 14 January 1994.

Kotz, Samuel and Johnson, Norman L. (Editors). *Breakthroughs in Statistics*. Volume II. Springer-Verlag, New York. 1993.

Kratovil, Robert. *Real Estate Law*. Prentice Hall, Englewood Cliffs. 1952.

Krishnamurti, J. *From Darkness to Light*. Harper & Row, Publishers, San Francisco. 1980.

Kruskal, William. 'Coordination Today: A Disease or a Disgrace' in *The American Statistician*. Volume 37, Number 3. 1983.

Kruskal, William. 'Statistics, Molière, and Henry Adams' in *American Scientist Magazine*. Volume 55. 1967.

Krutch, Joseph Wood. *Human Nature and the Human Condition*. Random House, New York. 1959.

Kyburg, Jr., H.E. and Smokler, H.E. (Editors). *Studies in Subjective Probability*. Wiley, New York. 1964.

Lang, Andrew. *Lost Leaders*. Longman, Green, and Co., New York. 1889.

Lapin, Lawrence L. *Statistics for Modern Business Decisions*. Harcourt Brace Jovanovich, Inc., New York. 1973.

Laplace, Pierre-Simon. *Essai Philosophique sur les Probabilités*. Courcier, Paris. 1814.

Laplace, Pierre-Simon. *Philosophical Essay on Probabilities*. Translated from the fifth French edition of 1825 by Andrew I Dale. Springer-Verlag New York, Inc., New York. 1995.

Laterius, Diogenes. *The Lives of Eminent Philosophers*. Harvard University Press, Cambridge. 1958.

Laut, Agnes. *The Conquest of the Great Northwest*. Musson Book Co., Toronto. 1908.

Leacock, Stephen. *Literary Lapses*. John Lane Co., New York. 1914.

Leacock, Stephen. 'Mathematics for Golfers' in *The World of Mathematics*. Volume 4. By James R. Newman. Simon and Schuster, New York. 1956.

LeCam, L. 'The Central Limit Theorem around 1935' in *Statistical Science*. Volume 1, Number 1. February 1986.

Lee, Hannah Farnham Sawyer. *The Log Cabin, or, The World Before You*. Appleton, Philadelphia. 1844.

Leibniz, Gottfried Wilhelm. *Leibniz: Discourses on Metaphysics*. Translated by George Montgomery. The Open Court Publishing Co., Chicago. 1902.

Leibniz, Gottfried Wilhelm. *Leibniz: Philosophical Papers and Letters*. Volume I. University of Chicago Press, Chicago. 1956.

LeSage, Alan René. *The Adventures of Gil Blas of Santillane*. Translated by Tobias Smollett. George Routledge and Sons, Ltd. London. 1881.

Lewis, Clarence Irving. *Mind and the World Order: Outline of a Theory of Knowledge*. Charles Scribner's Sons, New York. 1929.

Lewis, C.S. *The Pilgrim's Regress: An Allegorical Apology for Christianity, Reason and Romanticism*. Geoffrey Bles, London. 1933.

Lewis, C.S. *Christian Reflections*. Edited by Walter Hooper. Eerdmans, Grand Rapids. 1967.

Lewis, Don and Burke, C.J. 'The Use and Misuse of the Chi-Square Test' in *Psychological Bulletin*. Volume 46, Number 6. November 1949.

Lichtenberg, Georg. *Lichtenberg: Aphorisms & Letters*. Translated by Franz Mautner and Henry Hatfield. Jonathan Cape, London. 1959.

Lieber, Lillian R. *The Education of T.C. MITS*. W.W. Norton & Co., Inc., New York. 1944.

Lindley, Dennis V. 'Comment: A Tale of Two Wells' in *Statistical Science*. Volume 2, Number 1. February 1987.

Lipmann, Walter. *A Preface to Politics*. The University of Michigan Press, Ann Arbor. 1962.

Locke, John. *An Essay Concerning Human Understanding*. The Clarendon Press, Oxford. 1956.

Longair, M.S. 'Quasi-Stellar Radio Sources' in *Contemporary Physics*. Volume 8. 1967.

Longfellow, Henry Wadsworth. *The Poems of Longfellow*. Random House, Inc., New York. 1944.

Lonsdale, James and Lee, Samuel. *The Works of Virgil*. Macmillan and Co., London. 1883.

Lorenz, Konrad. *On Aggression*. Translated by Marjorie Kerr Wilson. Harcourt, Brace & World, Inc., New York. 1963.

Lover, S. *Rory O'More*. Little, Brown & Co., Boston. 1901.

Lucretius. *Lucretius on the Nature of Things*. Translated by Cyril Bailey. The Clarendon Press, Oxford. 1950.

Ludlum, Robert. *The Bourne Identity*. Richard Marek Publishers, New York. 1980.

Ludlum, Robert. *The Bourne Supremacy*. Random House, New York. 1986.

Ludlam, Robert. *The Parsifal Mosaic*. Random House, New York. 1982.

MacDonald, John D. *Condominium*. J.B. Lippincott Co., Philadelphia. 1977.

Macy, Arthur. *Poems*. W.B. Clarke & Co., Boston. 1905.

Maier, N.R.F. 'Maier's Law' in *The American Psychologist*. March 1960.

Malcolm, Andrew H. 'Data-Loving Japanese Rejoice on Statistics Day' in *The New York Times*. 26 October 1977.

Mallarmé, Stéphane. *Poems*. Translated by Roger Fry. Chatto & Windus, London. 1951.

Maloney, Russell. 'Inflexible Logic' in *The World of Mathematics*. Volume 4. By James R. Newman. Simon and Schuster, New York. 1956.

Manners, William. *Patience and Fortitude*. Harcourt Brace Jovanovich, New York. 1976.

Marlowe, Christopher. *Christopher Marlowe's Doctor Faustus*. Broadview Press, Peterborough. 1991.

Marlowe, Christopher. *Tamburlaine the Great. Part the First*. American Book Co., New York. 1912.

Martin, Thomas L., Jr. *Malice in Blunderland*. McGraw-Hill Book Co., New York. 1973.

Mason, Alpheus T. *Brandies: A Free Man's Life*. The Viking Press, New York. 1946.

Masters, Dexter. *The Accident*. Alfred A. Knopf, New York. 1955.

Mauldin, Bill. *Up Front*. Henry Holt and Co., New York. 1945.

May, R.M. 'Simple Mathematical Models with very Complicated Dynamics' in *Nature*. Volume 261. 1976.

McNemar, Quinn. 'Sampling in Psychological Research' in *Psychological Bulletin*. Volume 37, Number 6. June 1940.

Meitzen, August. *History, Theory, and Technique of Statistics*. Translated by Roland P. Falkner. American Academy of Political and Social Science. 1891.

Mellor, J.W. *Higher Mathematics for Students of Chemistry and Physics*. Dover Publications, New York. 1955.

Meredith, Owen. *Lucile*. Houghton Mifflin Co., Boston. 1882.

Metcalf, James J. *Poems*. Bantam Doubleday Dell Publishing Group, Inc., New York.

Meyer, Agnes. *Education for a New Morality*. Macmillan, New York. 1957.

Meyers, C.J., Jr. Discussion of E.G. Olds, 'On Some of the Essentials of the Control Chart Analysis' in *Transactions, American Society of Mechanical Engineers*. Volume 64. July 1942.

Michelmore, Peter. *Einstein: Profile of the Man*. Dodd, Mead & Co., New York. 1962.

Mikes, George. *How to be an Alien*. Basic Books, Inc., New York. 1964.

Miksch, W.F. 'The AVERAGE STATISTICIAN' in *Collier's*. 17 June 1950.

Mill, John Stuart. *Autobiography* in The Harvard Classics. Volume 25. P.F. Collier & Son Corp., New York. 1937.

Mill, John Stuart. *On Liberty*. Appleton–Century–Crofts, New York. 1947.

Mill, John Stuart. *System of Logic*. Longmans, Green, Reader & Dyer, London. 1868.

Miller, Henry. *Black Spring*. Grove Press, Inc., New York. 1963.

Milne, A.A. *Winnie-the-Pooh*. E.P. Dutton & Co., Inc., New York. 1961.

Milton, John. *Comus*. Charles Little and James Brown, Boston. 1845.

Milton, John. *Paradise Lost*. Cambridge University Press, Cambridge. 1972.

Milton, John. *Poetical Works of John Milton*. Porter and Coates, Philadelphia. No date.

Minnick, Wayne C. *The Art of Persuasion*. Houghton Mifflin Co., Boston. 1957.

Moger, Art. *The Complete Pun Book*. The Citadel Press, Secaucus. 1979.

Montaigne, Michel Eyquem de. *The Essays of Michel Eyquem de Montaigne*. Limited Edition Club, New York. 1946.

Moroney, M.J. *Facts from Figures*. Penguin Books. London. 1951.

Mosteller, F. 'Principles of Sampling' in *Journal of the American Statistical Association*. Volume 49, Number 265. 1954.

Mynors, R.A.B. *Collected Works of Erasmus*. Adages II vii 1 to III iii 100. University of Toronto Press, Toronto. 1992.

Nagi, S.Z. and R.G. Corwin. *The Social Contexts of Research*. John Wiley & Sons., Inc., New York. 1972.

Newton, Sir Isaac. *Mathematical Principles of Natural Philosophy*. Translated by Florian Cajori. University of California Press, Berkeley. 1960.

Newton, Sir Isaac. *Opticks*. Dover Publications, Inc., New York. 1952.

Nietzsche, Friedrich. 'The Joyful Wisdom' in *The Complete Works of Friedrich Nietzsche*. Volume 10. T.N. Foulis, Edinburgh. 1910.

Nightingale, Florence. *Notes on Nursing*. Appleton and Co., New York. 1860.

Nixon, Richard M. *The New York Times*. 10 November 1972.

Nizer, Louis. *My Life in Court*. Doubleday & Co., Inc., Garden City. 1944.

Nizer, Louis. *Thinking on Your Feet, adventures in speaking*. Liveright Publishing Corporation, New York. 1940.

Nye, Mary Jo. *Molecular Reality: A Perspective on the Scientific Work of Jean Perrin*. Macdonald, London. 1972.

Olds, Edwin G. and Knowler, Lloyd A. 'Teaching Statistical Quality Control for Town and Gown' in *Journal of the American Statistical Association*. Volume 44. 1949.

Oman, John. *The Natural and the Supernatural*. The Macmillan Co., New York. 1931.

O'Neil, W.M. *Fact and Theory*. Sydney University Press, Sydney. 1967.

Oppenheim, Abraham Naffali. *Questionnaire Design and Attitude Measurement*. Basic Books, New York. 1966.

Oppenheimer, Julius Robert. 'The Tree of Knowledge' in *Harper's Magazine*. Volume 217. October 1958.

O'Rielly, John. *In Bohemia*. The Pilot Publishing Co., Boston. 1886.

Orwell, George. *Nineteen Eighty-Four*. Harcourt, Brace & World, Inc., New York. 1949.

O'Shaughnessy, Arthur. *Ode*. Greenwood Press, Publishers, Westport. 1979.

Ovid. *Fasti*. Translated by Sir James George Frazer. Harvard University Press, London. 1959.

Ovid. *Metamorphoses*. Duke University Press, Durham. 1968.

Paley, William. *Paley's Natural Theology*. Harper & Brother Publisher, New York. 1855.

Papert, Seymour. *Mindstorms*. Basic Books, Inc., New York. 1980.

Parker, Tom. *Rules of Thumb*. Houghton Mifflin Co., Boston. 1983.

Parrish, Randall. *My Lady of the South*. A.C. McClurg & Co., Chicago. 1908.

Pascal, Blaise. *Pascal's Pensées*. Translated by H.F. Stewart. Pantheon Books Inc., New York. 1950.

Pascal, Blaise. *Scientific Treatise*. Encyclopedia B. 1952.

Pascal, Blaise. *The Thoughts of Blaise Pascal*. Greenwood Press, Publishers, Westport. 1975.

Paulos, John Allen. *Innumeracy*. Hill and Wang, New York. 1988.

Peacock, E.E. *Medical World News.* 1 September 1972.

Pearl, Judea. *Probabilistic Reasoning in Intelligent Systems.* Morgan Kaufman Publishing, San Mateo. 1988.

Pearson, E.S. 'The Choice of Statistical Test Illustrated on the Interpretation of Data Classed in a 2 × 2 Table' in *Biometrika.* Volume 34, Number 35. 1947.

Pearson, E.S. and Hartley, H.O. *Biometrika Tables for Statisticians.* Volume 1. Biometrika Trust, University College, London. 1984.

Pearson, Karl. *The History of Statistics in the 17th and 18th Centuries against the Changing Background of Intellectual, Scientific, and Religious Thought.* Macmillan Publishing Co., Inc., New York. 1978.

Pearson, Karl. 'Editorial' in *Biometrika.* 1901.

Peers, John. *1,001 Logical Laws.* Doubleday & Co., Inc., Garden City. 1979.

Peirce, Benjamin. 'Criterion for the Rejection of Doubtful Observations' in *The Astronomical Journal.* Number 45. 24 July 1852.

Peirce, Benjamin. 'Linear Associative Algebra' in *American Journal of Mathematics.* Volume 4. 1881.

Peirce, C.S. *Philosophical Writings of Peirce.* Selected and edited with an introduction by Justus Buchler. Dover Publications, Inc., New York. 1955.

Peirce, C.S. *Writings of Charles Sanders Peirce.* Volume 3, 1872–1878. Indiana University Press, Bloomington. 1958.

Penjer, Michael. *The New York Times.* 1 September 1989.

Perec, Georges. *Life, A User's Manual.* D.R. Godine, Boston. 1987.

Peter, Lawrence J. 'Peter's People' in *Human Behavior.* August 1976.

Petit, Jean-Pierre. *Euclid Rules OK?* Translated by Ian Stewart. John Murray (Publishers), Ltd, London. 1982.

Pettie, George. *A Petite Pallace of Pettie His Pleasures Containing Many Petie Histories by Him Set Forth in Comely Colours and Most Delightfully Discoursed.* Volume I. AMS Press, New York. 1970.

Pirandello, Luigi. *The Rules of the Game, The Life I Gave You [and] Lazarus.* Penguin Books, Middlesex. 1959.

Planck, Max. *A Survey of Physics.* Methuen & Co., Ltd., London. 1925.

Plato. *Crito.* Translated by B. Jowett. W.J. Black, New York. 1942.

Plato. *Gorgias.* Translated with notes by Terence Irwin. The Clarendon Press, Oxford. 1979.

Plato. *Phaedo.* Translated by B. Jowett. W.J. Black, New York. 1942.

Plato. *The Laws.* Translated with an introduction by A.E. Taylor. Dent Publishing, London. 1960.

Plato. *The Republic.* Translated with an Introduction by H.D.P. Lee. Penguin Books. 1955.

Plato. *Timaeus.* Translated by Francis M. Cornford. Liberal Arts Press, New York. 1959.

Plautus. *Aulularia.* Edited by E.G. Thomas. Clarendon Press, Oxford. 1913.

Playfair, William. *The Commercial and Political Atlas*. London. 1786.

Playfair, William. *The Statistical Breviary*. T. Bensley, London. 1801.

Plotinus. *The Six Enneads*. Translated by Stephen MacKenna. Larson Publications, Burdette. 1992.

Plutarch. *Plutarch's Lives*. Translation called Dryden's. Volume IV. Little, Brown, and Co., Boston. 1882.

Poe, Edgar Allen. *Tales of Mystery and Imagination*. Castle, New Jersey. 1985.

Pohl, Frederik. *The Coming of the Quantum Cats*. Bantam, New York. 1986.

Poincaré, Henri. *The Foundations of Science*. The Science Press, New York. 1913.

Polya, G. *Patterns of Plausible Inference*. Princeton University Press, New Jersey. 1954.

Polybius. *The Histories*. Harvard University Press, Cambridge. 1960.

Pomfret, John. *The Poetical Works of John Pomfret: with the Life of the Author*. The Apollo Press, Edinburgh. 1779.

Pope, Alexander. *The Complete Poetical Works of POPE*. Edited by Henry W. Boynton. Houghton Mifflin and Co., Boston. 1931.

Popper, Karl R. *Conjectures and Refutations*. Harper and Row, Publishers, New York. 1965.

Popper, Karl R. *The Logic of Scientific Discovery*. Basic Books, Inc., New York. 1959.

Popper, Karl R. *Realism and the Aim of Science*. Rowman & Littlefield, Ottawa. 1956.

Porter, Theodore M. *The Rise of Statistical Thinking*. Princeton University Press, Princeton. 1986.

Prakash, Satya. *Founders of Sciences in Ancient India*. The Research Institute of Ancient Scientific Studies, New Delhi. 1965.

Pratchett, Terry. *The Dark Side of the Sun*. St. Martin's Press, New York. 1976.

Price, Lucien. *Dialogues of Alfred North Whitehead*. Little, Brown and Co., Boston. 1954.

Prior, Matthew. *The Literary Works of Matthew Prior*. Volume I. Edited by H. Bunker Wright and Monroe K. Spears. The Clarendon Press, Oxford. 1959.

Proschan, Frank. 'Investigation of Latin Squares' in *Industrial Quality Control*. Volume XI, Number 1. July 1954.

Puzo, Mario. *Fools Die*. Putnam, New York. 1978.

Pynchon, Thomas. *Gravity's Rainbow*. The Viking Press, Inc., New York. 1973.

Pynchon, Thomas. *Slow Learner*. Little Brown & Co., Boston. 1984.

Queneau, Raymond. *Exercises in Style*. Translated by Barbara Wright. A New Directions Paperback, New York. 1981.

Quetelet, Adolphe. *Du système social et des lois qui le regissent*. Guillaumin, Paris. 1848.

Quetelet, Adolphe. *A Treatise on Man and the Development of His Faculties*. Scholar's Facsimiles & Reprints, Gainsville. 1969.

Rader, L.T. 'Putting Quality into Quantity' in *American Machinist*. Volume 87. 28 October 1943.

Raleigh, Walter. *Laughter from a Cloud*. Constable and Co., Ltd., London. 1923.

Ramsey, Frank Plumpton. *The Foundations of Mathematics and other Logical Essays*. Routledge & Kegan Paul Ltd., London. 1954.

Ramsey, James B. *Economic Forecasting: Models or Markets?* Sato Institute, San Francisco. 1980.

Ray, Donald P. *Trends in Social Science*. Philosophical Library, New York. 1961.

Read, Herbert. *Icon and Idea: The Function of Art in the Development of Human Consciousness*. Harvard University Press, Cambridge. 1955.

Reade, Charles. *A Terrible Temptation: a story of the day*. Chapman and Hall, London. 1871.

Redfield, Roy A. *Factors of Growth in a Law Practice*. Callaghan & Co., Mundelein. 1962.

Reichenbach, Hans. *The Rise of Scientific Philosophy*. University of California Press, Berkeley. 1953.

Reid, Thomas. *Essays on the Intellectual Power of Man*. Macmillan and Co., Limited, London. 1941.

Reynolds, Henry T. *Analysis of Nominal Data*. SAGE Publications, Beverly Hills. 1977.

Richardson, Samuel. *The History of Sir Charles Grandison, in a series of letters*. Routledge, London. No date.

Rickover, H.C. 'The World of the Uneducated' in *The Saturday Evening Post*. 28 November 1959.

Ritsos, Yannis. *Erotica*. Translated from the Greek by Kimon Friar. Sachm Press, Old Chatham. 1982.

Roberts, Nora. *Without a Trace*. MIRA Books, Ontario. 1990.

Robertson, John R. 'Transactions of the Statistical Society of London, Vol. 1, part 1' in *Westminster Review*. Volume 29. 1838.

Robertson, Louis Newton. *History and Organization of Criminal Statistics in the United States*. Patterson Smith, Montclair. 1969.

Roche, James Jeffrey. *Life of John Boyle O'Reilly*. The Mershon Co., New York. 1891.

Rogers, Will. *The Will Rogers Book*. Compiled by Paula McSpadden Love. Texian Press, Waco. 1972.

Rogers, Will. *The Writings of Will Rogers*. Volume IV. Oklahoma State University Press, Stillwater. 1980.

Rohault, Jacques. *Rohault's System of Natural Philosophy*. Johnson Reprint Corporation, New York. 1969.

Romanoff, Alexis L. *Encyclopedia of Thought*. Ithaca Heritage Books, Ithaca. 1975.

Ross, JoAnn. *Tempting Fate*. MIRA Books, Ontario. 1987.

Rousseau, Jean Jacques. *The Social Contract*. Translated by G.D.H. Cole. E.P. Dutton & Co., New York. 1950.

Royce, Joshiah. *The World and the Individual*. Dover Publications, Inc., New York. 1959.

Rudner, R. 'Remarks on Value Judgment in Scientific Validation' in *Scientific Monthly*. Volume 79. September 1954.

Rumanoff, Alexis L. *Encyclopedia of Thoughts, aphorisms, couplets and epigrams*. Ithaca Heritage Books, Ithaca. 1975.

Runyon, Damon. 'A Nice Place' in *Collier's*. 8 September 1934.

Russell, Bertrand. *Nightmares of Eminent Persons*. Simon and Schuster, New York. 1955.

Russell, Bertrand. *Principles of Mathematics*. W.W. Norton & Co., Inc., New York. 1938.

Russell, Bertrand. *The Analysis of Matter*. Dover Press, New York. 1954.

Russell, Bertrand. *The Scientific Outlook*. W.W. Norton & Co., Inc., New York. 1931.

Russell, E.J. 'Field Experiments: How They are Made and What They Are' in *Journal of the Ministry of Agriculture of Great Britain*. Volume 32. 1926.

Salsburg, David S. 'The Religion of Statistics as Practiced in Medical Journals' in *The American Statistician*. Volume 39, Number 3. August 1985.

Samuelson, Paul A. 'Science and Stocks' in *Newsweek*. 19 September 1966.

Santayana, George. *The Life of Reason*. Charles Scribner's Sons, New York. 1953.

Sarton, George. *Sarton on the History of Science*. Harvard University Press., Cambridge. 1962.

Sartre, Jean-Paul. *The Philosophy of Existentialism*. Philosophical Library, Inc., New York. 1965.

Savage, L.J. *The Foundations of Statistics*. John Wiley & Sons, New York. 1954.

Sayers, Dorothy L. *The Unpleasantness at the Bellona Club*. Harper & Row, New York. 1956.

Schlozer, Ludwig. *Westminster Review*. Volume I, Part I. 1838.

Schiller, Friedrich von. *Wallenstein: A Historical Drama in Three Parts*. Translated by Charles E. Possage. Frederick Unger Publishing Co., New York. 1958.

Schopenhauer, Arthur. *Parerga and Paralipomena: Short Philosophical Essays*. Translated by E.F.J. Payne. The Clarendon Press, Oxford. 1974.

Schumacher, E.F. *Good Work*. Harper & Row, New York. 1979.

Scott, Sir Walter. *The Fortunes of Nigel*. Adam and Charles Black, London. 1898.

Seaton, G.L. 'The Statistician and Modern Management' in *The American Statistician*. Volume 2, Number 6. December 1948.

Seeger, Raymond J. *Journal of the Washington Academy of Sciences*. Volume 36. 1946.

Segal, Erich. *Man, Woman and Child*. Harper and Row, Publishers. New York. 1980.

Seldes, George. *The Great Quotations*. A Caesar–Stuart Book, Lyle Stuart, New York. 1960.

Seuss, Dr. *The Cat in the Hat*. Houghton Mifflin, Boston. 1957.

Shaffer, Peter. *Two Plays by Peter Shaffer*. Athenaeum, New York. 1974.

Shapere, Dudly. *Philosophical Problems of Natural Science*. The Macmillan Co., New York. 1965.

Shapiro, Karl. *Collected Poems 1940–1978*. Random House, New York. 1978.

Shaw, George Bernard. *Back to Methuselah*. Brentans, New York. 1921.

Shaw, George Bernard. 'Great Catherine' in *Complete Plays with Prefaces*. Volume IV. Dodd, Mead & Co., New York. 1963.

Shaw, George Bernard. 'The Vice of Gambling and the Virtue of Insurance' in *The World of Mathematics*. Volume 3. By James R. Newman. Simon and Schuster, New York. 1956.

Shelley, Percy Bysshe. *The Poems of Percy Bysshe Shelley*. Methuen and Co., Ltd., London. 1911.

Sherman, Susan. *With Anger/With Love*. Mulch Press, Amherst. 1974.

Shewhart, W.A. 'Contributions of Statistics to the Science of Engineering' in *University of Pennsylvania Bicentennial Conference. Volume on Fluid Mechanics and Statistical Methods*. University of Pennsylvania Press, Philadelphia. 1941.

Simon, Herbert. *Models of Man: Social and Rational*. Wiley, New York. 1957.

Simpson, Thomas. 'A Letter to the Right Honorable George Earl of Macclesfield, President of the Royal Society, on the Advantage of Taking the Mean of a Number of Observations, in Practical Astronomy' in *Philosophical Transactions of the Royal Society of London*. Volume 49. 1755.

Skinner, B.F. *Walden Two*. Macmillan Publishing Co., Inc., New York. 1976.

Slonim, Morris James. *Sampling*. Simon and Schuster, New York. 1960.

Smedley, Frank. *Frank Fairlegh*. A. Hall, Virtue, and Co., London. 1850.

Smith, Logan. *Trivia*. Doubleday Press, New York. 1917.

Smith, Reginald H. 'A Sequel: The Bar is Not Overcrowded' in *American Bar Association Journal*. Volume 45, September 1959.

Smollett, Tobias. *The Life and Adventures of Sir Launcelot Greaves*. Oxford University Press, London. 1973.

Snedecor, G.W. 'On a Unique Feature of Statistics' in *Journal of the American Statistical Association*. Volume 44, Number 245. March 1949.

Snedecor, G.W. *Statistical Papers in Honor of George W. Snedecor*. Edited by T.A. Bancroft. The Iowa State University Press, Ames. 1972.

Sophocles. *The Plays of Sophocles*. Translated by Thomas Franklin. G. Routledge & Sons, London. 1893.

Spearman, Charles. *Psychology Down the Ages*. Volume I. Macmillan and Co., Ltd., London. 1937.

Spencer-Brown, George. *Probability and Scientific Inference*. Longmans, Green, London. 1957.

Stamaty, Mark Alan. 'Washingtoon' in *Time*. 25 September 1995.

Stamp, Josiah. *Some Economic Factors in Modern Life*. P.S. King & Son, Ltd., Orchard House. 1929.

Steadman, Frank M. 'Quality Control Posts Mill-Production Odds' in *Textile World*. Volume 94. Jul–Dec 1944.

Stekel, Wilhelm. *Marriage at the Crossroads*. Translated by Allen D. Gorman. W. Godwin, Inc., New York. 1931.

Sterne, Laurence. *Tristram Shandy*. J.M. Dent & Sons, Ltd., London. 1964.

Stewart, Alan. 'Averages' in *Times*. 4 January 1954.

Stewart, Ian. *Does God Play Dice?* Basil Blackwell Inc., Cambridge. 1990.

Stigler, Stephen M. *The History of Statistics: The Measurement of Uncertainty before 1900*. The Belknap Press of Harvard University Press, Cambridge. 1986.

Stone, Irving. *Clarence Darrow for the Defense*. Doubleday, Garden City. 1975.

Stoppard, Tom. *Night and Day*. Faber and Faber Ltd, London. 1978.

Stoppard, Tom. *Rosencrantz and Guildenstern are Dead*. Grove Press, Inc. New York. 1967.

Stout, Rex. *Death of a Doxy*. The Viking Press, New York. 1966.

Streatfield, Geoffrey. 'Sayings of the Week' in *The Observer*. 19 March 1950.

Strong, Lydia. 'Sales Forecasting: Problems and Prospects' in *Management Review*. September 1956.

Strunsky, Simeon. *Topics of the Times*. 30 November 1944.

Suidas. *Collected Works of Erasmus*. Translated and Annotated by R.A.B. Mynars. University of Toronto Press, Toronto. 1992.

Swift, Jonathan. *Gulliver's Travels*. Rinehart, New York. 1948.

Swift, Jonathan. *Satires and Personal Writings*. Oxford University Press, New York. 1932.

Swift, Jonathan. *The Portable Swift*. Edited by Carl Van Doren. The Viking Press, New York. 1966.

Sylvester, J.J. *Philosophical Magazine*. Volume 24. 1844.

Szilard, Leo. *Leo Szilard: His Version of the Facts: Selected Recollections & Correspondence*. Edited by Spencer R. Weart & Gertrude Weiss Szilard. The MIT Press, Cambridge. 1978.

Tanur, Judith, Mosteller, Frederick, Kruskal, William H., Lehmann, Erich L., Link, Richard F., Piters, Richard S. and Rising, Gerald R. *Statistics: A Guide to the Unknown*. Wadsworth Inc., Belmont. 1989.

Tarbell, Ida M. *The Ways of Woman*. The Macmillan Co., New York. 1916.

Tchekhov, Anton. *Tchekhov's Plays and Stories*. Translated by S.S. Koteliansky. E.P. Dutton and Co., Inc., New York. 1962.

Tennyson, Alfred Lord. *The Poems and Plays of Alfred Lord Tennyson*. The Modern Library, New York. 1938.

Terence. *Adelphoe*. Eldradge, Philadelphia. 1874.

Thackery, William M. *The Books of Snobs; and Sketches and Travels in London*. Smith Elder, London. 1869.

The Editors. 'Statistics, The Physical Sciences and Engineering' in *The American Statistician*. Volume 11, Number 4. August 1948.

The Editors. 'The Statistician and Everyday Affairs' in *The American Statistician*. Volume 11, Number 5. 1948.

The RAND Corporation. *A Million Random Digits with 100,000 Normal Deviates*. The Free Press, Publishers, Glenco. 1955.

Thiery, Paul Henri, Baron d'Holbach. *The System of Nature*. Volume I. Garland Publishing, Inc., New York. 1984.

Thompson, D'Arcy. *On Growth and Form*. Volume I. Cambridge University Press, London. 1959.

Thompson, William (Lord Kelvin). *Popular Lectures and Addresses*. Macmillan and Co., London. 1891.

Thomsett, Michael C. *The Little Black Book of Business Statistics*. American Management Association, New York. 1990.

Thoreau, Henry David. *Walden*. Bramhall House, New York. 1970.

Thoreau, Henry David. *Winter*. Houghton, Mifflin Co., Boston. 1888.

Thorn, John and Palmer, Peter. *The Hidden Game of Baseball*. Doubleday, New York. 1983.

Thucydides. *Thucydides: The History of the Peloponnesian War*. E.D. Dutton, New York. 1910.

Thurber, James. *Further Fables for Our Time*. Simon and Schuster, New York. 1956.

Thurber, James. *Lanterns and Lances*. Harper Publishing, New York. 1961.

Thurston, L.L. 'Current Issues in Factor Analysis' in *Psychological Bulletin*. Volume 37. April 1940.

Tippett, L.C. 'Sampling and the Standard Error' in *The World of Mathematics*. Volume 3. By James R. Neuman. Simon and Schuster, New York. 1956.

Tippett, L.H.C. *The Method of Statistics*. Williams and Norgate, Ltd. 1931.

Toffler, Alvin. *Future Shock*. Random House, New York. 1970.

Tolstoy, Leo. *War and Peace*. Carlton House, New York. No date.

Trollope, Anthony. *The Eustace Diamond*. Oxford University Press, Oxford. 1983.

Tsu, Chuang. *Inner Chapters*. Translated by Gia-Fu Feng and Jane English. Alfred A. Knopf, New York. 1974.

Tsu, Lao. *Tao Te Ching*. Translated by Gia-Fu Feng and Jane English. Alfred A. Knopf, New York. 1974.

Tufte, Edward R. *The Visual Display of Quantitative Information*. Graphics Press, Connecticut. 1983.

Tukey, John W. 'Statistical and Quantitative Methodology' in *Trends in Social Science*. Edited by Donald P. Ray. Philosophical Library, New York. 1961.

Tukey, John W. 'The Future of Data Analysis' in *Annals of Mathematical Statistics*. Volume 33, Number 1. March 1962.

Tukey, J.W. 'We Need both Exploratory and Confirmatory' in *The American Statistician*. Volume 34. 1980.

Tukey, John W. 'Where Do We Go From Here?' in *Journal of the American Statistical Association*. Volume 55, Number 268. March 1960.

Tukey, John W. 'Unsolved Problems of Experimental Statistics' in *Journal of the American Statistical Association*. Volume 49, Number 268. December 1954.

Turgenev, Ivan. *Father and Sons*. Translated by Alexandria Tolstoy. Bantam Books, New York. 1981.

Twain, Mark. *Adam's Diary*. Harper's Magazine. Volume 102, Number 611. April 1901.

Twain, Mark. *Huckleberry Finn*. Clarkson & Potter, Inc., New York. 1981.

Twain, Mark. *Mark Twain Laughing*. University of Tennessee Press, Knoxville. 1985.

Twain, Mark. *Pudd'nhead Wilson*. Harper & Brothers Publishers, New York. 1899.

Twain, Mark. *The Autobiography of Mark Twain*. Edited by Charles Neider. Harper & Row, Publishers, New York. 1959.

Unknown. *Adventures of Sylvia Hughes*. Garland Publishing, Inc., New York. 1975.

Van der Post, Laurens. *A Far off Place*. Hogarth Press, London. 1974.

Venn, J. 'On the Nature and Uses of Averages' in *Journal of the Royal Statistical Society*. Volume 54. 1891.

Venn, J. *The Logic of Chance*. Macmillan, London. 1888.

Villon, François. *The Poems of François Villon*. Translated by H.B. McCaskie. Cresset Press, London. 1946.

Volkart, Edmund H. *The Angel's Dictionary*. Franklin Watts, Inc., New York. 1986.

Voltaire. 'Philosophical Dictionary' in *The Portable Voltaire*. The Viking Press, New York. 1965.

Voltaire. 'Candide' in *Candide and Other Writings*. Random House, Inc. 1956.

von Clausewitz, Karl. *On War*. Edited and translated by Michael Howard and Peter Paret. Princeton University Press, Princeton. 1976.

von Mises, Richard. *Mathematical Theory of Probability and Statistics*. Edited by Hilda Geiringer. Academic Press, New York. 1964.

von Mises, Richard. *Probability, Statistics and Truth*. Academic Press, New York. 1964.

Walcott, Derek. *Collected Poems*. Farrar, Strauss & Giroux, New York. 1986.

Walker, Marshall. *The Nature of Scientific Thought*. Prentice-Hall, Inc., New Jersey. 1963.

Waller, Robert. *The Bridges of Madison County*. Warner Books, New York. 1992.

Wallis, W.A. 'The Statistical Research Group, 1942–1945' in *Journal of the American Statistical Association*. Volume 75, Number 370. June 1980.

Walsh, John E. *Handbook of NonParametric Statistics*. Volume I. Van Nostrand Co., Inc., Princeton. 1962.

Walton, Izaak. *The Complete Angler*. T.N. Foulis, London. 1913.

Wang, Chamont. *Sense and Nonsense of Statistical Inference: Controversy, Misuse, and Subtlety*. Marcel Dekker, Inc., New York. 1993.

Waugh, Evelyn. *The Letters of Evelyn Waugh*. Edited by Mark Amory. Ticknor & Fields, New York. 1980.

Wellman, Francis. *The Art of Cross-Examination*. Macmillan Publishing Co., New York. 1924.

Wells, H.G. *Mankind in the Making*. Chapman & Hall, London. 1904.

Wells, H.G. *The Work, Wealth and Happiness of Mankind*. William Heinemann, Ltd., London. 1932.

West, Jessamyn. *The Quaker Reader*. The Viking Press, New York. 1962.

West, Nathaniel. *Miss Lonelyhearts*. Farrar, Strauss and Giroux, New York. 1971.

Weyl, Hermann. *The Theory of Groups and Quantum Mechanics*. Dover Publications, Inc., New York. 1950.

White, E.B. *The Trumpet of the Swan*. Harper & Row, New York. 1970.

Whitehead, Alfred North. *Adventures of Ideas*. The Free Press, New York. 1967.

Whitehead, Alfred North. *An Introduction to Mathematics*. Oxford University Press, London. 1972.

Whitehead, Alfred North. *Process and Reality*. The Humanities Press, New York. 1929.

Whitehead, Alfred North. *Science and the Modern World*. The Macmillan Co., New York. 1967.

Whyte, Lancelot Law. *Essays on Atomism: from Democritus to 1960*. Wesleyan University Press, Middletown. 1961.

Wigner, Eugene P. 'The Unreasonable Effectiveness of Mathematics in the Natural Sciences' in *Communications in Pure and Applied Mathematics*. Volume 13. 1960.

Wilde, Oscar. *Epigrams: Phrases and Philosophies for the Use of the Young.* A.R. Keller & Co., Inc. New York. 1907.

Wilde, Oscar. *Oscar Wilde's Plays, Prose Writings, and Poems.* J.M. Dent & Sons, Ltd., London. 1966.

Wilde, Oscar. *The Importance of Being Earnest: A Trivial Comedy for Serious People.* William Heineman, London. 1949.

Wilde, Oscar. *The Picture of Dorian Gray.* The World Publishing Co., Cleveland. 1946.

Wilder, Thornton. *The Eighth Day.* Harper & Row, Publishers, New York. 1967.

Wilkins, John. *Of the Principles & Duties of Natural Religion.* Johnson Reprint Corp., New York. 1969.

Williams, Monier. *The Story of Nala.* The Clarendon Press, Oxford. No date.

Wilson, E.B. *Bulletin of the American Mathematical Association.* Volume 18. 1912.

Wittgenstein, Ludwig. *Tractatus Logico-Philosophicus.* Translated by D.F. Pears and B.F. McGuinness. Humanities Press Inc., New York. 1961.

Wolfowitz, J. 'Reflections on the Future of Mathematical Statistics' in *Essays in Probability and Statistics.* Edited by R.C. Bose *et al.* University of North Carolina Press, Chapel Hill. 1969.

Wonnacott, Ronald J. and Wonnacott, Thomas H. *Introductory Statistics.* John Wiley & Sons, Inc., New York. 1969.

Woodward, Robert. *Probability and Theory of Errors.* John Wiley & Sons, Inc., New York. 1906.

Wordsworth, William. *Wordsworth's Poetry and Prose.* The Clarendon Press, Oxford. 1921.

Wright, William Aldis. *The Complete Works of William Shakespeare.* Doubleday Doran & Co., Inc. 1936.

Wright, Jim. 'Second Thought' in *The Dallas Morning News.* 9 September 1969.

Yates, F. '(Discussion)' in *Journal of the Royal Statistical Society.* Series B, Volume 17. 1955.

Yeates, W.B. *The Collected Poems of W.B. Yeats.* The Macmillan Co., New York. 1961.

Youden, W.J. *Experimentation and Measurement.* National Science Teachers Association, Washington, D.C. 1962.

Yule, George U. *An Introduction to the Theory of Statistics.* Hafner Pub. Co., New York. 1950.

Yule, G.U. 'On the Theory of Correlation' in *Journal of the Royal Statistical Association.* December 1897.

PERMISSIONS

Grateful acknowledgement is made to the following for their kind permission to reprint copyright material. Every effort has been made to trace copyright ownership but if, inadvertently, any mistake or omission has occurred, full apologies are herewith tendered.

Full references to authors and the titles of their works are given under the appropriate quotation.

1,001 LOGICAL LAWS by John Peers. Copyright 1979. Reprinted by permission of the publisher, Doubleday & Company. Garden City, New Jersey.

AN ESSAY CONCERNING HUMAN UNDERSTANDING by John Locke. Copyright 1956. Reprinted by permission of the publisher, The Clarendon Press. Oxford, UK.

AN INTRODUCTION TO MATHEMATICS by Alfred North Whitehead. Copyright 1972. Reprinted by permission of the publisher, Oxford University Press. Oxford, UK.

AN INTRODUCTION TO PROBABILITY THEORY AND ITS APPLICA-TIONS by William Feller. Copyright 1960. Reprinted by permission of the publisher, John Wiley & Sons, Inc. New York.

A PHILOSOPHICAL ESSAY ON PROBABILITIES by Marquis de Pierre Simon Laplace. Copyright 1995. Reprinted by permission of the publisher, Springer-Verlag New York, Inc. New York.

ASTRONOMY TRANSFORMED by David O. Edge and Michael J. Mulkay. Copyright 1976. Reprinted by permission of the publisher, John Wiley & Sons, Inc. New York.

BLACK SPRING by Henry Miller. Copyright 1963. Reprinted by permission of the publisher, Grove Press, Inc. New York.

BRAIN 2000 by Earnest K. Gann. Copyright 1980. Reprinted by permission of the publisher, Doubleday & Company, Inc. Garden City, New Jersey.

BREAKTHROUGHS IN STATISTICS edited by Samuel Kotz and Norman L. Johnson. Copyright 1993. Reprinted by permission of the publisher, Springer-Verlag New York, Inc. New York.

BURNS POETICAL WORKS edited by J. Logie Robertson. Copyright 1966. Reprinted by permission of the publisher, Oxford University Press. Oxford, UK.

CAMILLA by Frances Burney. Copyright 1972. Reprinted by permission of the publisher, Oxford University Press. Oxford, UK.

CAROLINA CHANSONS by DuBose Heywood. Copyright 1922. Reprinted by permission of the publisher, Macmillan Publishing Company. New York.

CHANCE, SKILL, AND LUCK by John Cohen. Copyright 1960. Reprinted by permission of the publisher, Penguin Books. Baltimore, Maryland.

CHARLES BOOTH'S LONDON by Charles Booth. Copyright 1971. Reprinted by permission of the publisher, Penguin Books Ltd. London, UK.

CLARENCE DARROW FOR THE DEFENSE by Irving Stone. Copyright 1941. Reprinted by permission of the publisher, Doubleday, Doran & Company, Inc. Garden City, New Jersey.

COLLECTED POEMS 1940–1978 by Karl Shapiro. Copyright 1986. Reprinted by permission of the publisher, Wiesner & Wiesner, Inc. New York.

COMPLETE SPEAKER'S AND TOASTMASTER'S LIBRARY by Jacob M. Braude. Copyright 1967. Reprinted by permission of the publisher, Prentice Hall/A Division of Simon & Schuster. New York.

COMPLETE WRITINGS by William Blake. Copyright 1972. Reprinted by permission of the publisher, Oxford University Press. Oxford, UK.

'Consolidating our Gains' by Willford King. Copyright 1936. Reprinted by permission of the publishers, *Journal of the American Statistical Association*.

COWPER: POETICAL WORKS by William Cowper. Copyright 1967. Reprinted by permission of the publisher, Oxford University Press. Oxford, UK.

DIALOGUES OF ALFRED NORTH WHITEHEAD by Lucien Price. Copyright 1954. Reprinted by permission of the publisher, Little, Brown and Company. Boston, Massachusetts.

'Discussion' by G.E.P. Box. Copyright 1956. Reprinted by permission of *Journal of the Royal Statistical Society*.

THE MATHEMATICAL APPROACH TO BIOLOGY AND MEDICINE by Norman Bailey. Copyright 1967. Reprinted by permission of the publisher, John Wiley & Sons. Chichester, UK.

THE MILL ON THE FLOSS by George Eliot. Copyright 1980. Reprinted by permission of the publisher, The Clarendon Press. Oxford, UK.

THE NATURE OF SCIENCE AND OTHER LECTURES by Edwin Powell Hubble. Copyright 1977. Reprinted by permission of The Huntington. San Marino, California.

THE PARSIFAL MOSAIC by Robert Ludlum. Copyright 1982. Reprinted by permission of the publisher, Random House, Inc. New York.

THE PHYSICAL PRINCIPLES OF QUANTUM THEORY by Werner Heisenberg. Reprinted by permission of the publisher, Dover Publications, Inc. New York.

THE PROCESS OF EDUCATION by Jerome S. Bruner. Copyright 1960. Reprinted by permission of the publisher, The Belknap Press of Harvard University Press. Cambridge, Massachusetts.

THE ROMAN WAY by Edith Hamilton. Copyright 1932. Reprinted by permission of the publisher, W.W. Norton & Company, Inc. New York.

THE SMALL BACK ROOM by Nigel Balchin. Copyright 1943. Reprinted by permission of the publisher, Collins Publishing. London, UK.

THE SOCIAL CONTEXTS OF RESEARCH by S.Z. Nagi and R.G. Corwin. Copyright 1972. Reprinted by permission of the publisher, John Wiley and Sons. New York.

THE STORY OF NALA by Monier Williams. Reprinted by permission of the publisher, The Clarendon Press. Oxford, UK.

THE THEORY OF GROUPS AND QUANTUM MECHANICS by Hermann Weyl. Copyright 1950. Reprinted by permission of the publisher, Dover Publications, Inc. New York.

'The Whole Duty of the Statistical Forecaster' by Robert W. Burgess. Copyright 1937. Reprinted by permission of the publisher, *Journal of the American Statistical Association.*

THE WORLD AND THE INDIVIDUAL by Josiah Royce. Copyright 1959. Reprinted by permission of the publisher, Dover Publications, Inc. New York.

UP FRONT by Bill Mauldin. Copyright 1945. Reprinted by permission of the publisher, Henry Holt and Company. New York.

'Unsolved Problems of Experimental Statistics' by John Tukey. Copyright 1954. Reprinted by permission of the publisher, *Journal of the American Statistical Association.*

SUBJECT BY AUTHOR INDEX

...the question "How many legs
   does a normal man have?"
   should be answered by
   finding a statistical average,
   12

averages
Bernard, Claude
   Another very frequent
      application to biology is the
      use of averages..., 7
Bowley, Arthur L.
   Great numbers and the averages
      resulting from them..., 7
Brandies, Louis D.
   I abhor averages, 7
Froude, James Anthony
   We have to consider the million,
      not the units, 9
Juster, Norton
   But averages aren't real...they're
      just imaginary, 12
Nightingale, Florence
   ...an inveterate habit of taking
      averages..., 13
Venn, J.
   Why do we resort to averages at
      all?, 17

averages, chemical
Bernard, Claude
   Chemical averages are also often
      used, 7

averages, law of
Keegan, John
   ...'hitting the target',...is
      henceforth to be left to the
      law of averages, 12
Snood, Grover
   "You can't fight the law of
      averages,"..., 14
Stewart, Alan
   The only remedy would seem
      to be to repeal the law of
      averages, 15
Stoppard, Tom

The law of averages,...,means
   that if six monkeys were
   thrown up in the air..., 15

averages, statistical
Russell, Bertrand A.
   Scientific laws...are always,
      at least in physics, either
      differential equations, or
      statistical averages, 14

-B-

**Babylonical Statistical
   Association**
King, Willford
   ...clay tablets recording the
      minutes of the 1242 annual
      meeting of the Babylonical
      Statistical Association, 214

Bayesians
Bartlett, M.S.
   Bayesians should also take care
      to distinguish their various
      denominations..., 18
Luchenbruch, Peter
   There might also be some
      specialized terminology for
      Bayesians, but I have not
      seen any, 18
Wang, Chamont
   ...there are at least 46,656
      varieties of Bayesians, 18

Bernoulli
Harris, Sidney
   Which Bernoulli do you wish to
      see?, 72
Kelly-Bootle, Stan
   A mathematician in Reno..., 72

Bernoulli's theorem
Kneale, W.
   A misunderstanding of
      Bernoulli's theorem is
      responsible for one of the
      commonest fallacies in the
      estimation of probabilities,
      72

best estimate
Durand, David
...an estimate having optimum
qualities..., 60
Binary
Pynchon, Thomas
But a hardon, that's either there,
or it isn't, 74
biometrician
Fleiss, Joseph L.
There was a biometrician named
Mabel,..., 225
biostatistician
Unknown
A biostatistician talks statistics
to the biologist..., 231
blunders
Hugo, Victor
Great blunders are often make,
like large ropes, of a
multitude of fibers, 79

-C-
causalities
Disraeli, Benjamin
But great things spring from
causalities, 21
causality
Russell, Bertrand
But we are not likely to find
science returning to the
crude form of causality
believed in by Fijians..., 27
Unknown
I am not a heretic, 28
causation
Heise, David R.
Causation depends on an
extraordinary turning of
reality..., 23
cause
Akenside, Mark
Give me to learn each secret
cause..., 19
Bergson, Henri

...what is found in the effect was
already in the cause, 20
Buddhist Maxim
Every effect becomes a cause, 20
Da Vinci, Leonardo
There is no result in nature
without a cause..., 21
De Spinoza, Benedict
By CAUSE of itself, I understand
that, whose essence involves
existence..., 21
From a given determinate
cause an effect necessarily
follows..., 21
Froude, James Anthony
Every effect has its cause, 23
Holmes, O.W.
But he who, blind to universal
laws, Sees but effects,
unconscious of the cause—,
23
Hume, David
...every effect is a distinct event
from its cause, 24
It is universally allowed that
nothing exists without a
cause..., 24
Matthew
Wherefore by their fruits ye
shall know them, 28
Mill, John Stuart
...every fact which has a
beginning has a cause..., 25
Ovid
The cause is hidden..., 26
Pascal, Blaise
They saw the thing, but not the
cause, 26
Pettie, George
Sutch as the cause of everything
is, sutch wilbe the effect, 26
Plotinus
On the assumption that all
happens by Cause..., 26
Polybius

When we look at the plants
and bushes clothing an
entangled bank, we are
tempted to attribute their
proportional numbers and
kinds to what we call
chance, 38

De Moivre, Abraham
...some of the Problems about
Chance having a great
appearance of Simplicity,
the Mind is easily drawn
into a belief, that their
Solution may be attained by
the meer Strength of natural
good Sense, 38

Democritus
Nothing can come into being
from that which is not, or
pass away into what is not,
38

Dryden, John
...Chance rules all above..., 39

Eddington, Sir Arthur Stanley
...I shall certainly hit on a tune,
39

Eldridge, Paul
Value depends upon price and
price upon chance and
caprice, 39

Euripides
...it is chance that rules the
mortal sphere, 39

Galsworthy, John
It's all chance, but we can't stop
now, 40

Guest, Judith
...it is chance and not perfection
that rules the world, 40

Helvetius, C.A.
...chance, that is, an infinite
number of events, with
respect to which our
ignorance will not permit us
to perceive their causes..., 41

If chance be generally
acknowledged to be the
author of most discoveries
in almost all the arts..., 41

Homer
Therefore turn, and charge at
the foe, to stand or fall as is
the game of war..., 41

Hume, David
Though there be no such thing
as *Chance* in the world..., 41

Johnson, Samuel
Nothing was ever said with
uncommon felicity, but by
the cooperation of chance...,
42

Jonah
...and the lot fell upon Jonah, 48

Longfellow, Henry Wadsworth
I shot an arrow into the air,
It fell to earth I know not
where..., 42

Masters, Dexter
I should estimate...that there is
one chance in ten nothing
will happen with the
bomb..., 42
There wasn't more
than one chance in
God-knows-what..., 42

Milton, John
Chance governs all, 43
The power which erring men
call chance, 42

Nietzsche, Friedrich
No conqueror believes in
chance, 43

Paley, William
There must be *chance* in the
midst of design, 43

Pascal, Blaise
A game is being played at the
extremity of this infinite
distance where heads or
tails will turn up..., 43

...such a thing existed as an "Index of Correlation"..., 54

correlation, laws of
Galton, Francis
...there is a vast field of topics that fall under the laws of correlation..., 54

correlation, statistical
Dickson, Paul
There is a statistical correlation between the number of initials in an Englishman's name and his social class..., 53

correlational method
Cronbach, L.J.
The correlational method, for its part, can study what man has not learned to control, 53

correlations
Pearson, Karl
Biological phenomena in their numerous phases, economical and social, were seen to be only differentiated from the physical by the intensity of their correlations, 54

counted
Enarson, Harold L.
It does not follow that because something *can* be counted it therefore *should* be counted, 56

critical ratio
Unknown
The critical ratio is Z-ness..., 219

curve
Unknown
When you get an 8 on the midterm, there ain't a curve in the world that can save you, 74

-D-

data
Berkeley, Edmund E.
There is no substitute for honest, thorough, scientific effort to get correct data..., 55
Deming, William Edwards
Anyone can easily misuse good data, 55
Scientific data are not taken for museum purposes..., 56
There is only one kind of whiskey, but two broad classes of data, good and bad, 55
Durand, David
Deified numbers, 60
Ehrenberg, A.S.C.
Data are often presented in a form that is not immediately clear, 56
Fisher, Sir Walter A.
No human mind is capable of grasping in its entirety the meaning of any considerable quantity of numerical data, 56
Freeman, R. Austin
I can only suggest that, as we are practically without data, we should endeavor to obtain some, 56
Galton, Francis
My data were very lax..., 56
Holmes, Sherlock
...it is an error to argue in front of your data, 56
Data! Data! Data!, 57
It is a capital mistake to theorize before one has the data, 57
No data yet..., 57
Hooke, Robert
If you can't have an experiment, do the best you can with

If an experiment works,
something has gone wrong,
83
The experiment may be
considered a success if..., 83
Browning, Robert
Just an experiment for candour's
sake, 83
Cahier, Charles
Experiment is the mother of
science, 83
Carroll, Lewis
This is a most interesting
experiment..., 83–4
Cox, Gertrude
The statistician who supposes
that his main contribution
to the planning of an
experiment will involve
statistical theory..., 84
Darwin, Charles
If you knew some of the
experiments...which I am
trying, you would have a
good right to sneer..., 84
Eldridge, Paul
Those who fear muddy feet will
never discover new paths,
85
Emerson, Ralph Waldo
All life is an experiment, 85
Fisher, Sir Ronald A.
Every experiment may be said
to exist only in order to
give the facts a chance
of disproving the null
hypothesis, 85
He can perhaps say what the
experiment died of, 85
Godwin, William
No experiment can be more
precarious than that of a
half-confidence, 85
Green, Celia
...experiment is measurement, 61

Hume, David
...it being justly esteemed an
unpardonable temerity to
judge the whole course
of nature from one single
experiment..., 85
Hunter, John
Why think? Why not try the
experiment?, 86
Huxley, Thomas H.
Ancient traditions, when tested
by the severe process of
modern investigation..., 86
Jefferson, Thomas
...in the full tide of successful
experiment..., 86
Kapitza, Pyetr Leonidovich
...theory is a good thing but
a good experiment lasts
forever, 86
Kendall, Maurice G.
Hiawatha Designs an
Experiment, 86
Paracelsus, Philippus Aureolus
Every experiment is like a
weapon..., 86
Planck, Max
If one wishes to obtain a definite
answer from nature one
must attack the question
from a more general and
less selfish point of view, 86
Poincaré, Henri
Experiment is the sole source of
truth, 87
Rutherford, Ernest
If your experiment needs
statistics..., 87
The Bible
Prove all things, 87
Unknown
Diversity of treatment has been
responsible for much of the
criticism leveled against the
experiment, 88

Information that is imperfectly acquired, is generally as imperfectly retained..., 270

innumerancy

Paulos, John Allen

...an inability to deal comfortably with the fundamental notions of numbers..., 62

interviewer

Deutscher, I.

...neither the interviewer nor the instrument should act in any way upon the situation, 266

-J-

jackknife

Tukey, John W.

TEACH them the JACKKNIFE, 219

-K-

knowledge

Fischer, Robert B.

...the scientist must recognize the statistical aspect of much of his knowledge..., 126

Jevons, W.S.

We have...to content ourselves with partial knowledge..., 126

Skinner, B.F.

We give them an excellent survey of the methods and techniques of thinking..., 126

Sophocles

Nay, Knowledge must come through action..., 141

knowledge, common

Barry, Frederick

Common knowledge is...nothing else than the raw material which...has served as the basic substance of its vastly elaborated synthesis, 59

-L-

labeling

Eliot, George

The mere fact of naming an object tends to give definiteness to our conception of it..., 3

Latin Square

Kendall, Maurice G.

The first mathematical discussion of the Latin Square known to modern statisticians was given by Euler in 1882, 214

law

Bloch, Arthur

Negative expectations yield negative results, 127

Huxley, Thomas H.

Law means a rule..., 130

Krass, F.

...the law of measure and numbers rules in the changeful hosts of the stars as it does in man's thinking brain, 130–31

Rhodes, Charles E.

When any principle, law, tenet..., 132

Whitehead, Alfred North

If the law states a precise result, almost certainly it is not precisely accurate..., 133

law of averages

Boulle, Pierre

It's not a question of training, but the law of averages, 127

Coates, Robert M.

...the Law of Averages had never been incorporated into the body of federal jurisprudence..., 128

Mauldin, Bill

I feel like a fugitive from th' law of averages, 131

It was a desperate strategy,
    based on probabilities..., 174
Masters, Dexter
    ...the only procedure consistent
    with man's development
    was to follow where the
    probabilities led, 174
Pearl, Judea
    Probabilities are summaries of
    knowledge..., 175
Peirce, Charles Sanders
    ...it may be doubtful if there
    is a single extensive
    treatise on probabilities in
    existence which does not
    contain solutions absolutely
    indefensible, 175
Plato
    ...these arguments from
    probabilities are impostors...,
    176
Proverb, Italian
    A thousand probabilities does
    not make one fact, 178
Reade, Charles
    I feign probabilities, 178
Sartre, Jean-Paul
    When we want something, we
    always have to reckon with
    probabilities, 178
Schiller, Friedrich
    It is better to be satisfied
    with probabilities than to
    demand impossibilities and
    starve, 179
Voltaire
    Almost all human life depends
    on probabilities, 180
    He who has heard the
    thing told by twelve
    thousand eye-witnesses,
    has only twelve thousand
    probabilities..., 180
Von Clausewitz, Karl

...there is an interplay of
    possibilities, probabilities,
    good luck and bad that
    weaves its way throughout
    the length and breadth of
    the tapestry, 180
probabilities, calculus of
Arago
    The calculus of
    probabilities...ought to
    interest...the mathematician,
    the experimentalist, and the
    statesman, 159
Poincaré, Henri
    The very name of calculus of
    probabilities is a paradox,
    176
probabilities, theory of
Woodward, Robert S.
    The theory of probabilities and
    the theory of errors now
    constitute a formidable body
    of knowledge..., 182
probabilitiz
Billings, Josh
    The probabilitiz that the abuv
    probabilitz will assimilate
    themselfs tew the principal
    probabilitiz in the case, 160
probability
Adams, Douglas
    Probability factor one to one...,
    158
Arbuthnot, John
    I believe the Calculation of the
    Quantity of Probability
    might be improved to a
    very useful and pleasant
    Speculation..., 158
Aristotle
    A Probability is a thing that
    usually happens..., 159
Arnould, Antoine
    To judge what one must do
    to obtain a good or avoid

There is one habit which is clearly of British origin—that of queueing, 190

-R-

random

Carroll, Lewis
Couldn't put Humpty Dumpty in his place again, 191

Peers, John
Random stomping seldom catches bugs, 191

Sophocles
'Tis best to live at random..., 192

The RAND Corporation
A Million Random Digits with 100,000 Normal Deviates, 191

Unknown
Random is not haphazard, 192

random normal deviate

Durand, David
A contradiction in terms..., 61

random numbers

Heinlein, Robert A.
That kitten doesn't have a brain, he just has a skull full of random numbers..., 191

randomness

Brown, Spencer
The concept of randomness arises partly from games of chance, 195

Cohen, John
...nothing is so alien to the human mind as the idea of randomness, 191

reason

Bernard, Claude
To learn, we must necessarily reason about what we have observed..., 193

Bierce, Ambrose
To weigh probabilities in the scales of desire, 60

Eldridge, Paul
Reason is the shepherd trying to corral life's vast flock of wild irrationalities, 193

Holmes, Sherlock
...the grand thing is to be able to reason backward, 194

You fail...to reason from what you see, 193

Newton, Sir Isaac
My design in this book is...to propose and prove by reason and experiments..., 194

reasoning

Beveridge, W.I.B.
How easy it is for unverified assumption to creep into our reasoning unnoticed!, 193

Minnick, Wayne C.
This kind of reasoning has weakness..., 194

Romanoff, Alexis L.
Reasoning goes beyond the analysis of facts, 194

Watson, Dr.
Like all Holmes' reasoning the thing seemed simplicity itself when it was once explained, 195

reasons

Kasner, Edward
...principle of insufficient reasons, 194

Shakespeare, William
His reasons are as two grains of wheat hid in two bushels of chaff..., 195

recurse

Unknown
To iterate is human, to recurse divine, 196

The stormy holiday roads had yielded more than the statistical expectation of traffic accidents, 218

statistical generalizations

Lewis, Clarence Irving

Let us call these last "statistical generalizations" since they are exhibited at their best when supported by statistical procedures, 214

statistical graphics

Fienberg, Stephen E.

...we have no theory of statistical graphics..., 210

Tufte, Edward R.

...statistical graphics...are only as good as what goes into them, 115

Excellence in statistical graphics consists of complex ideas communicated with clarity, precision, and efficiency, 115

statistical illiterates

Pynchon, Thomas

Why am I surrounded by statistical illiterates?, 217

statistical improbability

Dawkins, Richard

The essence of life is statistical improbability on a colossal scale, 208

statistical inference

Wang, Chamont

...the whole notion of "statistical inference" often is more of a plague and less of a blessing to research workers, 263

statistical inference, ten commandments of

Driscoll, Michael F.

The Ten Commandments of Statistical Inference, 209

statistical judgment

Meitzen, August

No statistical judgment deals with the unit..., 215

statistical knowledge

Playfair, William

Statistical knowledge...has not till within these last 50 years become a regular object of study, 217

statistical laws

Jones, Raymond F.

Statistical laws enable the insurance company to function..., 213

Pearson, Karl

There is much value in the idea of the ultimate laws being statistical laws, 216

statistical magic

Devons, Ely

Statistical magic...is a mystery to the public..., 209

statistical method

Bell, Eric T.

The statistical method is social mathematics par excellence, 206

To grasp and analyze mass-reactions...a mastery of the modern statistical method is essential, 206

Deming, William Edwards

The statistical method is more than an array of techniques, 208

Einstein, Albert

By applying the statistical method we cannot foretell the behavior of an individual in a crowd, 210

Lippmann, Walter

The statistical method is of use only to those who have found it out, 215

Pearson, E.S.

History has never regarded itself as a science of statistics, 234

Taking for granted that the alternative to art was arithmetic, he plunged deep into statistics..., 234

Advertisement

...and you thought "impressive" statistics were 36-24-36, 234

Angell, Roger

Statistics are the food of love, 235

Balchin, Nigel

Probably knows no statistics whatever, 235

Bartlett, M.S.

It is concerned with things we can count, 235

Baudrillard, Jean

Like dreams, statistics are a form of wish fulfillment, 235

Belloc, Hilaire

...statistics come under the heading of lying..., 235

Before the curse of statistics fell upon mankind we lived a happy, innocent life..., 235

Statistics are the triumph of the quantitative method..., 235

Bernard, Claude

As for statistics, they are given a great role in medicine..., 236

Boorstin, Daniel J.

...statistics have tended to make facts into norms, 236

...statistics...displacers of moral imperatives..., 236

Booth, Charles

So far I speak only of impersonal statistics..., 236

Bowley, Arthur L.

...statistics deals with estimates..., 237

A knowledge of statistics is like a knowledge of foreign languages..., 237

Bowman, Scotty

Statistics are for losers, 237

Braude, Jacob M.

She was reading birth and death statistics..., 237

Browning, Elizabeth Barrett

We talk by aggregates, and think by systems and being used to face our evils in statistics..., 237

Burgess, Robert W.

The fundamental gospel of statistics is to push back the domain of ignorance..., 238

Burnman, Tom

...statistics, the term has no meaning unless the source, relevance, and truth are all checked out, 238

Byron, Lord

So that I do not grossly err in facts, Statistics, tactics, politics..., 238

Carlyle, Thomas

Statistics is a science which ought to be honourable..., 238

Statistics, one may hope, will improve gradually..., 238

Carroll, Lewis

...for he was weak in statistics..., 239

Coats, R.H.

...statistics has long been handmaid to these exact sciences..., 239

Cogswell, Theodore, R.

Statistics show that you have nothing to worry about, 239

Cohen, Jacob

...the typical behavioral scientist approaches applied

AUTHOR BY SUBJECT INDEX

-H-